IoT Edge Computing with MicroK8s

A hands-on approach to building, deploying, and distributing production-ready Kubernetes on IoT and Edge platforms

Karthikeyan Shanmugam

BIRMINGHAM—MUMBAI

IoT Edge Computing with MicroK8s

Group Product Manager: Rahul Nair
Publishing Product Manager: Surbhi Suman
Senior Editor: Shazeen Iqbal
Content Development Editor: Sujata Tripathi
Technical Editor: Rajat Sharma
Copy Editor: Safis Editing
Project Coordinator: Ashwin Dinesh Kharwa
Proofreader: Safis Editing
Indexer: Subalakshmi Govindhan
Production Designer: Aparna Bhagat
Marketing Coordinator: Nimisha Dua

First published: September 2022

Production reference: 2280922

Published by Packt Publishing Ltd.
Livery Place
35 Livery Street
Birmingham
B3 2PB, UK.

ISBN 978-1-80323-063-4

www.packt.com

To God Almighty, for this wonderful opportunity and for allowing me to complete it successfully. To my wife, Ramya, and to my daughters, Nethra and Kanishka, for being loving and supportive.

To my parents, Shanmugam and Jayabarathi.

Contributors

About the author

Karthikeyan Shanmugam is an experienced solutions architect professional, with about 20+ years of experience in the design and development of enterprise applications across various industries. Currently, he is working as a senior solutions architect at Amazon Web Services, where he is responsible for designing scalable, adaptable, and resilient architectures that solve client business challenges. Prior to that, he worked for companies such as Ramco Systems, Infosys, Cognizant, and HCL Technologies.

He specializes in cloud, cloud-native, containers, and container orchestration tools, such as Kubernetes, IoT, digital twin, and microservices domains, and has obtained multiple certifications from various cloud providers.

He is also contributing author in leading journals such as *InfoQ*, *Container Journal*, *DevOps.com*, *The New Stack*, and the **Cloud Native Computing Foundation** (**CNCF.io**) blog.

His articles on emerging technologies (including the cloud, Docker, Kubernetes, microservices, and cloud-native development) can be read on his blog at upnxtblog.com.

About the reviewers

Alex Chalkias is a senior product manager working with Kubernetes and cloud-native technologies, currently at Elastic. He was always most drawn to the intersection of business and technology, specifically aspiring to build amazing products and solve interesting problems using open source software. His professional background also includes Canonical, Amadeus, and Nokia, where he occupied the roles of software engineer, scrum master, business analyst, and product owner. Alex holds a master's degree in electrical engineering and computer science from the University of Patras. During his studies, he focused on programming and new technologies, such as augmented reality and human-computer interaction. In his spare time, he is an avid tennis, music, and TV series fan.

Jimmy Song is a developer advocate at Tetrate, a CNCF ambassador, and a cloud-native community (China) founder. He mainly focuses on cloud-native fields, including Kubernetes and service meshes. He is one of the authors of the books *Deeper Understanding of Istio* and *Future Architecture*.

Meha Bhalodiya is a final-year computer science engineering student. A Google Summer of Code 2022 scholar, she started contributing to Keptn's integration in automating deployment states after the state has been synced. In the spring of 2022, she was an LFX mentee for the CNCF-K8s Gateway API, where she contributed to assessing the project documentation, the contributor documentation, and the website. She was involved with Kubernetes 1.24 and 1.23 release team as a documentation shadow. She is also qualified as a **Linux Foundation Training (LiFT)** scholar. Additionally, at Kubernetes Community Days Bengaluru 2022, she got selected as a speaker and delivered a session on running local Kubernetes clusters using minikube, KinD, and MicroK8s.

Table of Contents

Part 1: Foundations of Kubernetes and MicroK8s

1

2

Part 2: Kubernetes as the Preferred Platform for IoT and Edge Computing

3

Essentials of IoT and Edge Computing   55

4

Handling the Kubernetes Platform for IoT and Edge Computing   67

Part 3: Running Applications on MicroK8s

5

Creating and Implementing Updates on a Multi-Node Raspberry Pi Kubernetes Clusters   81

6

Configuring Connectivity for Containers   111

7

Setting Up MetalLB and Ingress for Load Balancing   143

8

Monitoring the Health of Infrastructure and Applications   169

9

Using Kubeflow to Run AI/MLOps Workloads   203

10

Going Serverless with Knative and OpenFaaS Frameworks   235

Part 4: Deploying and Managing Applications on MicroK8s

11

Managing Storage Replication with OpenEBS   261

Preface

The idea for this book was born when one of my customers wanted to implement a minimal container orchestration engine for their apps on their resource-constrained edge device. Deploying the entirety of Kubernetes was not the solution, but then I encountered the realm of minimal Kubernetes distributions, and after much experimentation with several providers, I chose MicroK8s to successfully build various edge computing use cases and scenarios for them.

Canonical's MicroK8s Kubernetes distribution is small, lightweight, and fully conformant. It's a minimalistic distribution with a focus on performance and simplicity. MicroK8s can be easily deployed in IoT and edge devices due to its small footprint. By the end of this book, you will know how to effectively implement the following use cases and scenarios for edge computing using MicroK8s:

- Getting your Kubernetes cluster up and running

- Enabling core Kubernetes add-ons such as **Domain Name System (DNS)** and dashboards

- Creating, scaling, and performing rolling updates on multi-node Kubernetes clusters

- Working with various container networking options, such as Calico, Flannel, and Cilium

- Setting up MetalLB and Ingress options for load balancing

- Using OpenEBS storage replication for stateful application

- Configuring Kubeflow and running AI/ML use cases

- Configuring service mesh integration with Istio and Linkerd

- Running serverless applications using Knative and OpenFaaS

- Configuring logging and monitoring options (Prometheus, Grafana, Elastic, Fluentd, and Kibana)

- Configuring a multi-node, highly available Kubernetes cluster

- Configuring Kata for secured containers

- Configuring strict confinement for running in isolation

According to *Canonical's 2022 Kubernetes and cloud native operations report* (`https://juju.is/cloud-native-kubernetes-usage-report-2022`), 48 percent of respondents indicated the biggest barriers to migrating to or using Kubernetes and containers are a lack of in-house capabilities and limited staff.

As indicated in the report, there is a skills deficit as well as a knowledge gap, which I believe this book will solve by covering crucial areas that are required to bring you up to speed in no time.

Who this book is for

The book is intended for DevOps and cloud engineers, Kubernetes **Site Reliability Engineers (SREs)**, and application developers who desire to implement efficient techniques for deploying their software solutions. It will be also useful for technical architects and technology leaders who are looking to adopt cloud-native technologies. A basic understanding of container-based application design and development, virtual machines, networking, databases, and programming will be helpful to get the most out of this book.

What this book covers

Chapter 1, *Getting Started with Kubernetes*, introduces Kubernetes and the various components of the Kubernetes system as well as the abstractions.

Chapter 2, *Introducing MicroK8s*, introduces MicroK8s and shows how to install it, how to verify its installation status, and how to monitor and manage a Kubernetes cluster. We will also learn how to use some of the add-ons and deploy a sample application.

Chapter 3, *Essentials of IoT and Edge Computing*, delves into how Kubernetes, edge computing, and the cloud can collaborate to drive intelligent business decisions. This chapter gives an overview of the **Internet of Things (IoT)**, the Edge, and how they are related, as well as the advantages of edge computing.

Chapter 4, *Handling the Kubernetes Platform for IoT and Edge Computing*, examines how Kubernetes for edge computing offers a compelling value proposition and different architectural approaches that demonstrate how Kubernetes can be used for edge workloads, as well as support for architecture that meets an enterprise application's requirements – low latency, resource-constrained, data privacy, and bandwidth scalability.

Chapter 5, *Creating and Implementing on Updates Multi-Node Raspberry Pi Kubernetes Clusters*, explores how to set up a MicroK8s Raspberry Pi multi-node cluster, deploy a sample application, and execute rolling updates on the deployed application. We will also understand ways to scale the deployed application. We will also touch upon some of the recommended practices for building a scalable, secure, and highly optimized Kubernetes cluster model.

Chapter 6, *Configuring Connectivity for Containers*, looks at how networking is handled in a Kubernetes cluster. Furthermore, we will understand how to use Calico, Cilium, and Flannel CNI plugins to network the cluster. We will go through the most important factors to consider when choosing a CNI service.

Chapter 7, *Setting Up MetalLB and Ingress for Load Balancing*, delves into techniques (MetalLB and Ingress) for exposing services outside a cluster.

Chapter 8, Monitoring the Health of Infrastructure and Applications, examines various choices for monitoring, logging, and alerting your cluster, and provides detailed steps on how to configure them. We will also go through the essential metrics that should be watched in order to successfully manage your infrastructure and apps.

Chapter 9, Using Kubeflow to Run AI/MLOps Workloads, covers how to develop and deploy a sample ML model using the Kubeflow ML platform. We will also go through some of the best practices for running AI/ML workloads on Kubernetes.

Chapter 10, Going Serverless with Knative and OpenFaaS Frameworks, examines two of the most popular serverless frameworks included with MicroK8s, Knative and OpenFaaS, both of which are Kubernetes-based platforms for developing, deploying, and managing modern serverless workloads. We will also go through the best practices for developing and deploying serverless applications.

Chapter 11, Managing Storage Replication with OpenEBS, looks at how to use OpenEBS to implement storage replication that synchronizes data across several nodes. We will go through the steps involved in configuring and implementing a PostgreSQL stateful application utilizing the OpenEBS Jiva storage engine. We will also look at the Kubernetes storage best practices as well as recommendations for data engines.

Chapter 12, Implementing Service Mesh for Cross-Cutting Concerns, walks you through the steps of deploying Istio and Linkerd service meshes. You will also learn how to deploy and run a sample application, as well as how to configure and access dashboards.

Chapter 13, Resting Component Failure Using HA Clusters, walks you through the steps involved in setting up a highly available cluster that can withstand a component failure and continue to serve workloads without interruption. We will also discuss some of the best practices for implementing Kubernetes applications on your production-ready cluster.

Chapter 14, Hardware Virtualization for Securing Containers, looks at how to use Kata Containers, a secure container runtime, to provide stronger workload isolation, leveraging hardware virtualization technology. We also discuss the best practices for establishing container security on your production-grade cluster.

Chapter 15, Implementing Strict Confinement for Isolated Containers, shows you how to install the MicroK8s snap with a strict confinement option, monitor the installation's progress, and manage a Kubernetes cluster running on Ubuntu Core. We will also deploy a sample application and examine whether the application is able to run on a strict confinement-enabled Kubernetes cluster.

Chapter 16, Diving into the Future, looks at how Kubernetes and MicroK8s are uniquely positioned for accelerating IoT and edge deployments, and also the key trends that are shaping our new future.

Frequently Asked Questions About MicroK8s

To get the most out of this book

A basic understanding of container-based application design and development, virtual machines, networking, databases, and programming will be helpful to get the most out of this book. The following are the prerequisites for building a MicroK8s Kubernetes cluster:

Software/hardware covered in the book	Operating system requirements
<ul><li>A microSD card (4 GB minimum, with 8 GB recommended)</li><li>A computer with a microSD card drive</li><li>A Raspberry Pi 2, 3, or 4 (3 or more)</li><li>A micro-USB power cable (USB-C for the Pi 4)</li><li>A Wi-Fi network or an Ethernet cable with an internet connection</li><li>(Optional) A monitor with an HDMI interface</li><li>(Optional) An HDMI cable for the Pi 2 and 3 and a micro-HDMI cable for the Pi 4</li><li>(Optional) A USB keyboard</li><li>An SSH client such as PuTTY</li><li>A hypervisor such as Oracle VM VirtualBox 6.1 to create virtual machines</li></ul>	Windows or Linux to run Ubuntu virtual machines

If you are using the digital version of this book, we advise you to type the code yourself or access the code from the book. Doing so will help you avoid any potential errors related to the copying and pasting of code.

Download the example code files

You can download the example code files for this book from GitHub at `https://github.com/PacktPublishing/IoT-Edge-Computing-with-MicroK8s`. If there's an update to the code, it will be updated in the GitHub repository.

We also have other code bundles from our rich catalog of books and videos available at `https://github.com/PacktPublishing/`. Check them out!

Download the color images

We also provide a PDF file that has color images of the screenshots and diagrams used in this book. You can download it here: https://packt.link/HprZX.

Conventions used

There are a number of text conventions used throughout this book.

Code in text: Indicates code words in text, database table names, folder names, filenames, file extensions, pathnames, dummy URLs, user input, and Twitter handles. Here is an example: "To check a list of available and installed add-ons, use the status command."

A block of code is set as follows:

```
apiVersion: v1
kind: Service
metadata:
  name: metallb-load-balancer
spec:
  selector:
    app: whoami
  ports:
    - protocol: TCP
      port: 80
      targetPort: 80
  type: LoadBalancer
```

Any command-line input or output is written as follows:

```
kubectl apply -f loadbalancer.yaml
```

Bold: Indicates a new term, an important word, or words that you see onscreen. For instance, words in menus or dialog boxes appear in **bold**. Here is an example: "Navigate to **Monitoring** under **Namespaces** on the Kubernetes dashboard, and then click **Services**."

> **Tips or Important Notes**
> Appear like this.

Get in touch

Feedback from our readers is always welcome.

General feedback: If you have questions about any aspect of this book, email us at `customercare@packtpub.com` and mention the book title in the subject of your message.

Errata: Although we have taken every care to ensure the accuracy of our content, mistakes do happen. If you have found a mistake in this book, we would be grateful if you would report this to us. Please visit `www.packtpub.com/support/errata` and fill in the form.

Piracy: If you come across any illegal copies of our works in any form on the internet, we would be grateful if you would provide us with the location address or website name. Please contact us at `copyright@packt.com` with a link to the material.

If you are interested in becoming an author: If there is a topic that you have expertise in and you are interested in either writing or contributing to a book, please visit `authors.packtpub.com`.

Share Your Thoughts

Once you've read *IoT Edge Computing with MicroK8s*, we'd love to hear your thoughts! Scan the QR code below to go straight to the Amazon review page for this book and share your feedback.

https://packt.link/r/1803230630

Your review is important to us and the tech community and will help us make sure we're delivering excellent quality content.

Part 1: Foundations of Kubernetes and MicroK8s

In this part, you will be introduced to MicroK8s and its ecosystem. You will also learn how to install a MicroK8s Kubernetes cluster and get it up and running.

This part of the book comprises the following chapters:

- *Chapter 1, Getting Started with Kubernetes*
- *Chapter 2, Introducing MicroK8s*

1

Getting Started with Kubernetes

Kubernetes is an open source container orchestration engine that automates how container applications are deployed, scaled, and managed. Since it was first released 7 years ago, it has made great strides in a short period. It has previously had to compete with and outperform container orchestration engines such as Cloud Foundry Diego, CoreOS's Fleet, Docker Swarm, Kontena, HashiCorp's Nomad, Apache Mesos, Rancher's Cattle, Amazon ECS, and more. Kubernetes is now operating in an entirely different landscape. This indicates that developers only need to master one container orchestration engine so that they can be employed for 90% of container-related jobs.

The Kubernetes container orchestration framework is a ready-for-production open source platform built on Google's 15+ years of experience running production workloads, as well as community-contributed best-of-breed principles and concepts. Kubernetes divides an application's containers into logical units for easier administration and discovery. Containers (cgroups) have been around since early 2007 when they were first included in the mainline Linux kernel. A container's small size and portability allows it to host an exponentially higher number of containers than VMs, lowering infrastructure costs and allowing more programs to be deployed faster. However, until Docker (2013) came along, it didn't generate significant interest due to usability concerns.

Docker is different from standard virtualization; it is based on operating-system-level virtualization. Containers, unlike hypervisor virtualization, which uses an intermediation layer (hypervisor) to run virtual machines on physical hardware, run in user space on top of the kernel of an operating system. As a result, they're incredibly light and fast. This can be seen in the following diagram:

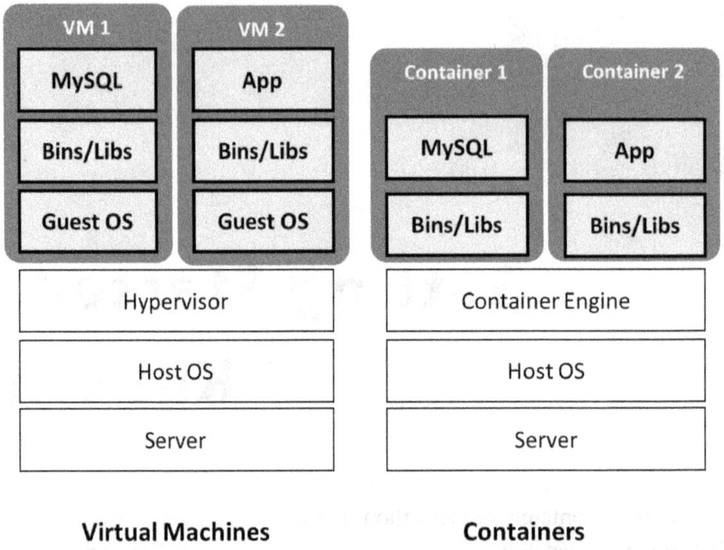

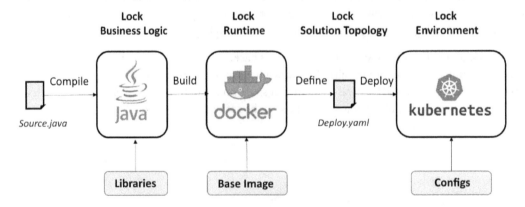

Figure 1.1 – Virtual machines versus containers

The Kubernetes container orchestration framework automates much of the operational effort that's necessary to run containerized workloads and services. This covers provisioning, deployment, scaling (up and down), networking, load balancing, and other tasks that software teams must perform to manage a container's life cycle. Some of the key benefits that Kubernetes brings to developers are as follows:

- **Declarative Application Topology**: This describes how each service should be implemented, as well as their reliance on other services and resource requirements. Because we have all of this data in an executable format, we can test the application's deployment parts early on in development and treat it like programmable application infrastructure:

Figure 1.2 – Declarative application topology

- **Declarative Service Deployments**: The update and rollback process for a set of containers is encapsulated, making it a repetitive and automated procedure.

- **Dynamically Placed Applications**: This allows applications to be deployed in a predictable sequence on the cluster, based on application requirements, resources available, and governing policies.

- **Flexible scheduler**: There is a lot of flexibility in terms of defining conditions for assigning pods to a specific or a set of worker nodes that meet those conditions.

- **Application Resilience**: Containers and management platforms help applications be more robust in a variety of ways, as follows:

  - Resource consumption policies such as CPU and memory quotas

  - Handling the failures using a circuit breaker, timeout, retry, and so on

  - Failover and service discovery

  - Autoscaling and self-healing

- **Self-Service Environments**: These allow teams and individuals to create secluded environments for CI/CD, experimentation, and testing purposes from the cluster in real time.

- **Service Discovery, Load Balancing, and Circuit Breaker**: Without the use of application agents, services can discover and consume other services. There's more to this than what is listed here.

In this chapter, we're going to cover the following main topics:

- The evolution of containers

- Kubernetes overview – understanding Kubernetes components

- Understanding pods

- Understanding deployments

- Understanding StatefulSets and DaemonSets

- Understanding jobs and CronJobs

- Understanding services

The evolution of containers

Container technology is a means of packaging an application so that it may run with separated dependencies, and its compartmentalization of a computer system has radically transformed software development today. In this section, we'll look at some of the key aspects, including where this technology originated and the background behind the container technology:

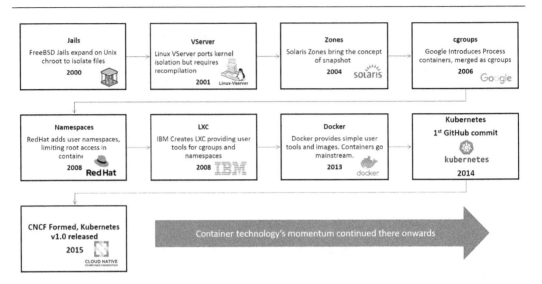

Figure 1.3 – A brief history of container technology

Early containers (chroot systems with Unix version 7), developed in the 1970s, offered an isolated environment in which services and applications could operate without interfering with other processes, thereby creating a sandbox for testing programs, services, and other processes. The original concept was to separate the workload of the container from that of production systems, allowing developers to test their apps and procedures on production hardware without disrupting other services. Containers have improved their abilities to isolate users, data, networking, and more throughout time.

With the release of Free BSD Jails in the 2000s, container technology finally gained traction. "Jails" are computer partitions that can have several jails/partitions on the same system. This jail architecture was developed in 2001 with Linux VServer, which included resource partitioning and was later linked to the Linux kernel with OpenVZ in 2005. Jails were merged with boundary separation to become Solaris Containers in 2004.

Container technology advanced substantially after the introduction of control groups in 2006. Control groups, or cgroups, were created to track and isolate resource utilization, such as CPU and memory. They were quickly adopted and improved upon in **Linux Containers** (**LXC**) in 2008, which was the most full and stable version of any container technology at the time since it functioned without changes having to be made to the Linux kernel. Many new technologies have sprung up because of LXC's reliability and stability, the first of which was Warden in 2011 and, more importantly, Docker in 2013.

Containers have gained a lot of usage since 2013 due to a slew of Linux distributions releasing new deployment and management tools. Containers running on Linux systems have been transformed into virtualization solutions at the operating system level, aiming to provide several isolated Linux environments on a single Linux host. Linux containers don't need their own guest operating systems; instead, they share the kernel of the host operating system. Containers spin up significantly faster than virtual machines since they don't require a specialized operating system.

Containers can employ Linux kernel technologies such as namespaces, Apparmor, SELinux profiles, chroot, and cgroups to create an isolated operational environment, while Linux security modules offer an extra degree of protection, ensuring that containers can't access the host machine or kernel. Containerization in terms of Linux provided even more versatility by allowing containers to run various Linux distributions from their host operating system if both were running on the same CPU architecture.

Linux containers provided us with a way to build container images based on a variety of Linux distributions, as well as an API for managing the containers' lifespan. Linux distributions also included client tools for dealing with the API, as well as snapshot features and support for moving container instances from one container host to another.

However, while containers running on a Linux platform broadened their applicability, they still faced several fundamental hurdles, including unified management, real portability, compatibility, and scaling control.

The emergence of Apache Mesos, Google Borg, and Facebook Tupperware, all of which provided varying degrees of container orchestration and cluster management capabilities, marked a significant advancement in the use of containers on Linux platforms. These platforms allowed hundreds of containers to be created instantly, and also provided support for automated failover and other mission-critical features that are required for container management at scale. However, it wasn't until Docker, a variation of containers, that the container revolution began in earnest.

Because of Docker's popularity, several management platforms have emerged, including Marathon, Kubernetes, Docker Swarm, and, more broadly, the DC/OS environment that Mesosphere built on top of Mesos to manage not only containers but also a wide range of legacy applications and data services written in, for example, Java. Even though each platform has its unique approach to orchestration and administration, they all share one goal: to make containers more mainstream in the workplace.

The momentum of container technology accelerated in 2017 with the launch of Kubernetes, a highly effective container orchestration solution. Kubernetes became the industry norm after being adopted by CNCF and receiving backing from Docker. Thus, using a combination of Kubernetes and other container tools became the industry standard.

With the release of cgroups v2 (Linux version 4.5), several new features have been added, including rootless containers, enhanced management, and, most crucially, the simplicity of cgroup controllers.

Container usage has exploded in the last few years (`https://juju.is/cloud-native-kubernetes-usage-report-2021`) in both emerging *"cloud-native"* apps and situations where IT organizations wish to "containerize" an existing legacy program to make it easier to lift and shift onto the cloud. Containers have now become the de facto standard for application delivery as acceptance of cloud-native development approaches mature.

We'll dive more into Kubernetes components in the next section.

Kubernetes overview – understanding Kubernetes components

In this section, we'll go through the various components of the Kubernetes system, as well as their abstractions.

The following diagram depicts the various components that are required for a fully functional Kubernetes cluster:

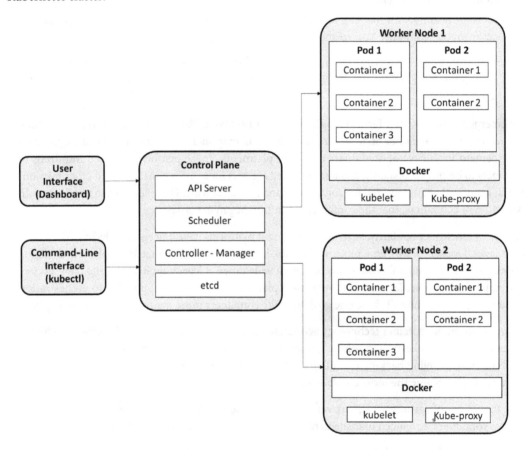

Figure 1.4 – A Kubernetes system and its abstractions

Let's describe the components of a Kubernetes cluster:

- *Nodes*, which are worker machines that run containerized work units, make up a Kubernetes cluster. Every cluster has at least one worker node.

- There is an API layer (Kubernetes API) that can communicate with Kubernetes clusters, which may be accessed via a command-line interface called *kubectl*.

There are two types of resources in a Kubernetes cluster (as shown in the preceding diagram):

- The control plane, which controls and manages the cluster
- The nodes, which are the workers' nodes that run applications

All the operations in your cluster are coordinated by the control plane, including application scheduling, maintaining the intended state of applications, scaling applications, and deploying new updates.

A cluster's nodes might be **virtual machines** (**VMs**) or physical computers that serve as worker machines. A kubelet is a node-managing agent that connects each of the nodes to Kubernetes control plane. Container management tools, such as Docker, should be present on the node as well.

The control plane executes a command to start the application containers whenever an application needs to be started on Kubernetes. Containers are scheduled to run on the cluster's nodes by the control plane.

The nodes connect to the control plane using the Kubernetes API that the control plane provides. The Kubernetes API allows end users to interface directly with the cluster. The master components offer the cluster's control plane capabilities.

API Server, Controller-Manager, and Scheduler are the three processes that make up the Kubernetes control plane. The Kubernetes API is exposed through the API Server. It is the Kubernetes control plane's frontend. Controller-Manager is in charge of the cluster's controllers, which are responsible for handling everyday activities. The Scheduler keeps an eye out for new pods that don't have a node assigned to them and assigns them one. Each worker node in the cluster is responsible for the following processes:

- **Kubelet**: This handles all the communication with the Kubernetes MasterControl plane.
- **kube-proxy**: This handles all the networking proxy services on each node.
- The container runtime, such as Docker.

Control plane components are in charge of making global cluster decisions (such as application scheduling), as well as monitoring and responding to cluster events. For clusters, there is a web-based Kubernetes dashboard. This allows users to administer and debug cluster-based applications, as well as the cluster itself. Kubernetes clusters may run on a wide range of platforms, including your laptop, cloud-hosted virtual machines, and bare-metal servers.

MicroK8s is a simplistic streamlined Kubernetes implementation that builds a Kubernetes cluster on your local workstation and deploys all the Kubernetes services on a tiny cluster that only includes one node. It can be used to experiment with your local Kubernetes setup. MicroK8s is compatible with Linux, macOS X, Raspberry Pi, and Windows and can be used to experiment with local Kubernetes setups or for edge production use cases. Start, stop, status, and delete are all basic bootstrapping procedures that are provided by the MicroK8s CLI for working with your cluster. We'll learn how to install MicroK8s, check the status of the installation, monitor and control the Kubernetes cluster, and deploy sample applications and add-ons in the next chapter.

Other objects that indicate the state of the system exist in addition to the components listed in *Figure 1.4*. The following are some of the most fundamental Kubernetes objects:

- Pods
- Deployments
- StatefulSets and DaemonSets
- Jobs and CronJobs
- Services

In the Kubernetes system, Kubernetes objects are persistent entities. These entities are used by Kubernetes to represent the state of your cluster. It will operate indefinitely to verify that the object exists once it has been created. You're simply telling the Kubernetes framework how your cluster's workloads should look by building an object; this is your cluster's ideal state. You must use the Kubernetes API to interact with Kubernetes objects, whether you want to create, update, or delete them. The CLI handles all Kubernetes API queries when you use the `kubectl` command-line interface, for example. You can also directly access the Kubernetes API in your apps by using any of the client libraries. The following diagram illustrates the various Kubernetes objects:

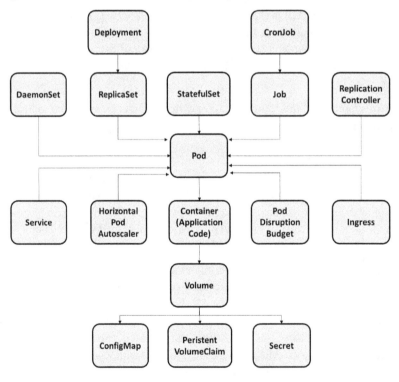

Figure 1.5 – Overview of Kubernetes objects

Kubernetes provides the preceding set of objects (such as pods, services, and controllers) to satisfy our application's requirements and drive its architecture. The guiding design principles and design patterns we employ to build any new services are determined by these new primitives and platform abilities. A *deployment* object, for example, is a Kubernetes object that can represent an application running on your cluster. When you build the *deployment*, you can indicate that three replicas of the application should be running in the *deployment* specification. The Kubernetes system parses the *deployment* specification and deploys three instances of your desired application, altering its status as needed. If any of those instances fail for whatever reason, the Kubernetes framework responds to the discrepancy between the specification and the status by correcting it – in this case, by establishing a new instance.

Understanding how Kubernetes works is essential, but understanding how to communicate with Kubernetes is just as important. We'll go over some of the ways to interact with a Kubernetes cluster in the next section.

Interacting with a Kubernetes cluster

In this section, we'll look at different ways to interface with a Kubernetes cluster.

Kubernetes Dashboard is a user interface that can be accessed via the web. It can be used to deploy containerized applications to a Kubernetes cluster, troubleshoot them, and control the cluster's resources. This dashboard can be used for a variety of purposes, including the following:

- All the nodes and persistent storage volumes are listed in the **Admin** overview, along with aggregated metrics for each node.

- The **Workloads** view displays a list of all running applications by namespace, as well as current pod memory utilization and the number of pods in a deployment that are currently ready.

- The **Discover** view displays a list of services that have been made public and have enabled cluster discovery.

- You can drill down through logs from containers that belong to a single pod using the **Logs** view.

- For each clustered application and all the Kubernetes resources running in the cluster, the **Storage** view identifies any persistent volume claims.

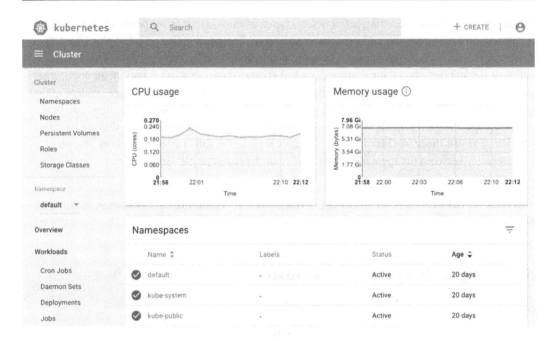

Figure 1.6 – Kubernetes Dashboard

- With the help of the Kubernetes command-line tool, `kubectl`, you can perform commands against Kubernetes clusters. `kubectl` is a command-line tool for deploying applications, inspecting and managing cluster resources, and viewing logs. `kubectl` can be installed on a variety of Linux, macOS, and Windows platforms.

  The basic syntax for `kubectl` looks as follows:

  ```
  kubectl [command] [type] [name] [flags]
  ```

  Let's look at `command`, `type`, `name`, and `flags` in more detail:

- `command`: This defines the action you wanted to obtain on one or more resources, such as `create`, `get`, `delete`, and `describe`.

- `type`: This defines the types of your resources, such as pods and jobs.

- `name`: This defines the name of the resource. Names are case-sensitive. If the name is omitted, details for all the resources are displayed; for example, `kubectl get pods`.

- `flags`: This defines optional flags.

We'll take a closer look at each of these Kubernetes objects in the upcoming sections.

Understanding pods

Pods are the minimal deployable computing units that can be built and managed in Kubernetes. They are made up of one or more containers that share storage and network resources, as well as running instructions. Pods have the following components:

- An exclusive IP address that enables them to converse with one another
- Persistent storage volumes based on the application's needs
- Configuration information that determines how a container should run

The following diagram shows the various components of a pod:

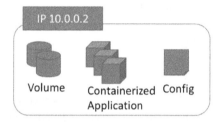

Figure 1.7 – The components of a pod

Workload resources known as controllers create pods and oversee the rollout, replication, and health of pods in the cluster.

The most common types of controllers are as follows:

- **Jobs** for batch-type jobs that are short-lived and will run a task to completion
- **Deployments** for applications that are stateless and persistent, such as web servers
- **StatefulSets** for applications that are both stateful and persistent, such as databases

These controllers build pods using configuration information from a pod template, and they guarantee that the operating pods meet the deployment specification provided in the pod template by creating replicas in the number of instances specified in the deployment.

As we mentioned previously, the Kubectl command-line interface includes various commands that allow users to build pods, deploy them, check on the status of operating pods, and delete pods that are no longer needed.

The following are the most commonly used `kubectl` commands concerning pods:

- The `create` command creates the pod:

  kubectl create -f FILENAME.

  For example, the `kubectl create -f ./mypod.yaml` command will create a new pod from the `mypod` YAML file.

- The `get pod/pods` command will display information about one or more resources. Information can be filtered using the respective label selectors:

  kubectl get pod pod1

- The `delete` command deletes the pod:

  kubectl delete -f FILENAME.

  For example, the `kubectl delete -f ./mypod.yaml` command will delete the `mypod` pod from the cluster.

With that, we've learned that a pod is the smallest unit of a Kubernetes application and is made up of one or more Linux containers. In the next section, we will look at deployments.

Understanding deployments

Deployment allows you to make declarative changes to pods and ReplicaSets. You can provide a desired state for the deployment, and the deployment controller will incrementally change the actual state to the desired state.

Deployments can be used to create new ReplicaSets or to replace existing deployments with new deployments. When a new version is ready to go live in production, the deployment can easily handle the upgrade with no downtime by using predefined rules. The following diagram shows an example of a deployment:

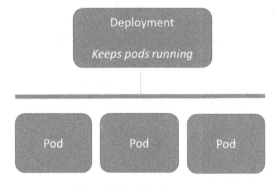

Figure 1.8 – A deployment

The following is an example of a deployment. It creates a ReplicaSet to bring up three `nginx` pods:

```
apiVersion: apps/v1
kind: Deployment
metadata:
  name: nginx-sample-deployment
  labels:
    app: nginx
spec:
  replicas: 3
  selector:
    matchLabels:
      app: nginx
  template:
    metadata:
      labels:
        app: nginx
    spec:
      containers:
      - name: nginx
        image: nginx:1:21
        ports:
        - containerPort: 80
```

In the preceding example, the following occurred:

- A deployment called `nginx-sample-deployment` is created, as indicated by the `metadata.name` field.

- The image for this deployment is set by the `Spec.containers.image` field (`nginx:latest`).

- The deployment creates three replicated pods, as indicated by the `replicas` field.

The most commonly used `kubectl` commands concerning deployment are as follows:

- The `apply` command creates the pod:

  kubectl apply -f FILENAME.

  For example, the `kubectl apply -f ./nginx-deployment.yaml` command will create a new deployment from the `nginx-deployment.yaml` YAML file.

- The `get deployments` command checks the status of the deployment:

```
kubectl get deployments
```

This will produce the following output:

```
NAME                      READY   UP-TO-DATE   AVAILABLE   AGE
nginx-sample-deployment   3/3        0             0
1s
```

The following fields are displayed:

- NAME indicates the names of the deployments in the namespace.

- READY shows how many replicas of the application are available.

- UP-TO-DATE shows the number of replicas that have been updated to achieve the desired state.

- AVAILABLE shows the number of available replicas.

- AGE indicates the length of time the application has been running.

- The `describe deployments` command indicates the details of the deployment:

```
kubectl describe deployments
```

- The `delete` command removes the deployment that was made by the `apply` command:

```
kubectl delete -f FILENAME.
```

With that, we have learned that deployments are used to define the life cycle of an application, including which container images to use, how many pods you should have, and how they should be updated. In the next section, we will look at StatefulSets and DaemonSets.

Understanding StatefulSets and DaemonSets

In this section, we'll go over two distinct approaches to deploying our application on Kubernetes: using StatefulSets and DaemonSets.

StatefulSets

The StatefulSet API object is used to handle stateful applications. A StatefulSet, like a deployment, handles pods that have the same container specification. A StatefulSet, unlike a deployment, continues using a persistent identity for each of its pods. These pods are generated for identical specifications, but they can't be exchanged: each has a unique identity that it keeps throughout any rescheduling.

The following example demonstrates the components of a StatefulSet:

```
apiVersion: v1
kind: Service
metadata:
  name: nginx
  labels:
    app: nginx
spec:
  ports:
  - port: 80
    name: web
  clusterIP: None
  selector:
    app: nginx
---
apiVersion: apps/v1
kind: StatefulSet
metadata:
  name: web
spec:
  selector:
    matchLabels:
      app: nginx
  serviceName: "nginx"
  replicas: 3
  template:
    metadata:
      labels:
        app: nginx
    spec:
      containers:
      - name: nginx
        image: nginx:latest
        ports:
        - containerPort: 80
```

```
            name: web
        volumeMounts:
        - name: www_volume
          mountPath: /usr/share/nginx/html
  volumeClaimTemplates:
  - metadata:
      name: www_volume
    spec:
      accessModes: [ "ReadWriteOnce" ]
      storageClassName: "my-storage-class"
      resources:
        requests:
          storage: 10Gi
```

In the preceding example, we have the following:

- `nginx` is the headless service that is used to control the network domain.

- web is the StatefulSet that has a specification that indicates that three replicas from the `nginx` container will be launched in unique pods.

- `volumeClaimTemplates` will use PersistentVolumes provisioned by a PersistentVolume provisioner to offer stable storage.

Now, let's move on to DaemonSets.

DaemonSets

A DaemonSet guarantees that all (or some) nodes have a copy of a pod running. As nodes are added to the cluster, pods are added to them. As nodes are removed from the cluster, garbage is collected in pods. When you delete a DaemonSet, the pods it produced are also deleted.

The following are some example use cases regarding DaemonSets:

- Run a daemon for cluster storage on each node, such as `glusterd` and `ceph`.

- Run a daemon for logs to be collected on each node, such as `Fluentd` or `FluentBit` and `logstash`.

- Run a daemon for monitoring on every node, such as Prometheus Node Exporter, `collectd`, or Datadog agent.

The following code shows a DaemonSet that's running the `fluent-bit` Docker image:

```
apiVersion: apps/v1
kind: DaemonSet
metadata:
  name: fluent-bit
  namespace: kube-system
  labels:
    k8s-app: fluent-bit
spec:
  selector:
    matchLabels:
      name: fluent-bit
  template:
    metadata:
      labels:
        name: fluent-bit
    spec:
      tolerations:
      - key: node-role.kubernetes.io/master
        operator: Exists
        effect: NoSchedule
      containers:
      - name: fluent-bit
        image: fluent/fluent-bit:latest
        resources:
          limits:
            memory: 200Mi
          requests:
            cpu: 100m
            memory: 200Mi
```

In the preceding example, the `fluent-bit` DaemonSet has a specification that tells `fluent-bit` to run on all the nodes.

The most commonly used `kubectl` commands concerning DaemonSets are as follows:

- The `create` or `apply` command creates the DaemonSet:

  `kubectl apply -f FILENAME.`

  For example, the `kubectl apply -f ./daemonset-deployment.yaml` command will create a new DaemonSet from the `daemonset-deployment.yaml` YAML file.

- The `get daemonset` command is used to monitor the status of the DaemonSet:

  `kubectl get daemonset`

  This will produce the following output:

  ```
  NAME                    READY   UP-TO-DATE   AVAILABLE   AGE
  daemonset-deployment    3/3        0            0
  1s
  ```

  The following fields are displayed:

  - `NAME` indicates the names of the DaemonSets in the namespace.

  - `READY` shows how many replicas of the application are available.

  - `UP-TO-DATE` shows the number of replicas that have been updated to achieve the desired state.

  - `AVAILABLE` shows how many replicas of the application are available.

- `AGE` indicates the length of time the application has been running.

- The `describe daemonset` command indicates the details of the DaemonSets:

  `kubectl describe daemonset`

- The `delete` command removes the deployment that was made by the `apply` command:

  `kubectl delete <<daemonset>>`

With that, we've learned that a DaemonSet ensures that all or a set of nodes run a copy of a pod, while a StatefulSet is used to manage stateful applications. In the next section, we will look at jobs and CronJobs.

Understanding jobs and CronJobs

In this section, we will learn how to use Kubernetes jobs to build temporary pods that do certain tasks. CronJobs are similar to jobs, but they run tasks according to a set schedule.

Jobs

A job launches one or more pods and continues to try executing them until a specific number of them succeed. The job keeps track of how many pods have been completed successfully. The task (that is, the job) is completed when a certain number of successful completions is met.

When you delete a job, it also deletes all the pods it created. Suspending a job causes all the current pods to be deleted until the job is resumed. The following code shows a job config that runs every minute and prints `example Job Pod is Running` as its output:

```
apiVersion: batch/v1
kind: Job
metadata:
  name: example-job
spec:
 template:
    spec:
      containers:
      - name: example-job
        image: busybox
        command: ['echo', 'echo example Job Pod is Running']
      restartPolicy: OnFailure
      backoffLimit: 4
```

The most commonly used `kubectl` commands concerning jobs are as follows:

- The `create` or `apply` command creates the pod:

  kubectl apply -f FILENAME.

  For example, the `kubectl apply -f ./jobs-deployment.yaml` command will create new jobs from the `jobs-deployment.yaml` YAML file.

- The `describe jobs` command indicates the details of the jobs:

  kubectl describe jobs <<job name>>

CronJob

A CronJob is a job that is created regularly. It is equivalent to a single line in a crontab (cron table) file. It executes a job that is written in Cron format regularly.

CronJobs are used to automate common processes such as backups and report generation. You can decide when the work should begin within that period by setting each of those jobs to repeat indefinitely (for example, once a day, week, or month).

The following is an example of a CronJob that prints the `example-cronjob Pod is Running` output every minute:

```
apiVersion: batch/v1
kind: CronJob
metadata:
  name: example-cronjob
spec:
  schedule: "*/1 * * * *"
  jobTemplate:
    spec:
      template:
        spec:
          containers:
          - name: example-cronjob
            image: busybox
            imagePullPolicy: IfNotPresent
            command:
            - /bin/sh
            - -c
            - date; echo example-cronjob Pod is Running ; sleep 5
          restartPolicy: OnFailure
```

Here, `schedule: /1 *` indicates that the crontab syntax is used in Linux systems.

Jobs and CronJobs are critical components of Kubernetes, particularly for performing batch processes and other critical ad hoc tasks. We'll examine service abstraction in the next section.

Understanding services

In Kubernetes, a **service** is an abstraction that defines a logical set of pods, as well as a policy for accessing them. An example service definition is shown in the following code block, which includes a collection of pods that each listen on TCP port `9876` with the `app=exampleApp` label:

```
apiVersion: v1
kind: Service
metadata:
```

```
      name: example-service
spec:
   selector:
      app: exampleApp
   ports:
      - protocol: TCP
        port: 80
        targetPort: 9876
```

In the preceding example, a new `Service` object named `example-service` was created that routes TCP port `9876` to any pod with the `app=exampleApp` label. This service is given an IP address by Kubernetes, which is utilized by the service proxies. A Kubernetes service, in simple terms, connects a group of pods to an abstracted service name and IP address. Discovery and routing between pods are provided by services. Services, for example, connect an application's frontend to its backend, which are both deployed in different cluster deployments. Labels and selectors are used by services to match pods with other applications.

The core attributes of a Kubernetes service are as follows:

- A label selector that locates pods
- The cluster IP address and the assigned port number
- Port definitions
- (Optional) Mapping for incoming ports to a targetPort

Kubernetes will automatically assign a cluster IP address, which will be used to route traffic by service proxies. The selector's controller will check for pods that match the defined label. Some applications will require multiple ports to be exposed via the service. Kubernetes facilitates this by using multi-port services, where a user can define multiple ports in a single service object.

In the following example, we have exposed ports `80` and `443` to target ports `8080` and `8090` to route HTTP and HTTPS traffic to any underlying pods using the `app=webserver-nginx-multiport-example` selector:

```
apiVersion: v1
kind: Service
metadata:
   name: nginx-service
spec:
   selector:
      app: webserver-nginx-multiport-example
```

```
ports:
  - name: http
    protocol: TCP
    port: 80
    targetPort: 8080
  - name: https
    protocol: TCP
    port: 443
    targetPort: 8090
```

A service can also be defined without the use of a selector; however, you must explicitly connect the service (IP address, port, and so on) using an endpoints object. This is because, unlike with a selector, Kubernetes does not know which pods the service should be connected to, so endpoint objects are not built automatically.

Some use cases for services without selectors are as follows:

- Connecting to another service in a different namespace or cluster

- Communicating with external services, data migration, testing services, deployments, and so on

Let's create a deployment with three replicas of an Apache web server:

```
apiVersion: apps/v1
kind: Deployment
metadata:
  name: apache-deployment
  labels:
    app: webserver
spec:
  replicas: 3
  selector:
    matchLabels:
      app: webserver
  template:
    metadata:
      labels:
        app: webserver
    spec:
      containers:
```

```
      - name: apache
        image: httpd:latest
        ports:
        - containerPort: 80
```

Create the deployment using the following command:

kubectl apply -f apache-deployment.yaml

The following are the most common types of services:

- **ClusterIP**: This is the default type and exposes the service via the cluster's internal IP address. These services are only accessible within the cluster. So, users need to implement port forwarding or a proxy to expose a ClusterIP to a wider ingress of traffic.

- **NodePort**: A static port on each node's IP is used to expose a service. To route traffic to the NordPort service, a ClusterIP service is automatically created. Requesting `NodeIP:NodePort>` from the outside allows users to communicate with the service.

- **LoadBalancer**: This is the preferred solution for exposing the cluster to the wider internet. The LoadBalancer type of service will create a load balancer (the load balancer's type depends on the cloud provider) and expose the service externally. It will also automatically create ClusterIP and NodePort services and route traffic accordingly.

- **ExternalName**: Maps a service to a predefined `externalName ex.sampleapp.test.com` field by returning a value for the CNAME record.

Summary

To conclude, Kubernetes is a container orchestration system that maintains a highly available cluster of machines that work together as a single entity. In this chapter, we discovered that Kubernetes supports several abstractions that allow containerized applications to be deployed to a cluster without being bound to specific machines. We also learned that pods represent a set of operating containers on your cluster. A deployment is an excellent fit for managing a stateless application workload on your cluster. StatefulSets can be used to run one or more connected pods to manage stateful applications, while DaemonSets specify pods and provide node-specific functionality. Finally, jobs and CronJobs handle batch processing and other key ad hoc tasks. In a nutshell, Kubernetes is a container orchestration system that is portable, extensible, and self-healing.

In the next chapter, we'll look at the lightweight Kubernetes engine known as MicroK8s, which can run on the edge, IoT, and appliances. MicroK8s is also ideal for offline prototyping, testing, and development.

2
Introducing MicroK8s

Kubernetes is the world's most popular orchestration technology for container-based applications, automating their deployment and scalability while also making maintenance easier. Kubernetes, on the other hand, has its own set of complications. So, how can an organization use containerization to address complexity while avoiding adding to Kubernetes' complexity?

Canonical's MicroK8s is a powerful Cloud-Native Computing Foundation-certified Kubernetes distribution. Here are some of the key reasons why it has become a powerful enterprise computing platform:

- **Delivered as snap packages**: These are application packages for desktop, cloud, and even **Internet of Things (IoT)** devices that are simple to install and secured with auto-updates, and they can be deployed on any of the Linux distributions that support snaps.

- **Strict confinement**: This ensures complete isolation from the underlying operating system as well as a highly secure Kubernetes environment fit for production.

- **Production-grade add-ons**: Add-ons such as Istio, Knative, CoreDNS, Prometheus, Jaeger, Linkerd, Cilium, and Helm are available. They are straightforward to set up, requiring only a few lines of commands. For better **Artificial Intelligence (AI)** and **Machine Learning (ML)** capabilities, Kubeflow is also available as an add-on to MicroK8s.

MicroK8s will speed up Kubernetes deployments due to its ability to decrease complexity. Treating devices like distributed containerized programs allows developers to concentrate on applications rather than infrastructure, making operations teams' lives easier.

MicroK8s enables you to combine Kubernetes installations into a single cluster and distribute workloads over one or more of these nodes. In this chapter, we are going to cover the following main topics:

- Introducing MicroK8s Kubernetes
- Quick installation
- Deploying a sample application
- Enabling add-ons

- Starting/stopping MicroK8s

- Configuring MicroK8s to use local images

- Configuring MicroK8s to use built-in registries

- Configuring MicroK8s to use private/public registries

- Configuring MicroK8s services

- Troubleshooting application and cluster issues

Introducing MicroK8s Kubernetes

MicroK8s is a production-ready Kubernetes distribution that is powerful, lightweight, and reliable. It's a Kubernetes distribution for enterprises with a reduced memory and disk footprint, as well as pre-installed add-ons such as Istio, Knative, Grafana, Cilium, and others. MicroK8s meets your needs, whether you're running a production setup or just getting started with Kubernetes.

Anyone who has attempted to work with Kubernetes understands how difficult it is to get up and running with the deployment. There are other minimalist solutions on the market that minimize deployment time and complexity, but they come at the cost of key extensibility and missing add-ons.

MicroK8s gets you up and running in just 60 seconds, so you don't waste too much time jumping through hurdles to get Kubernetes up and running.

Some of the key features are as follows:

- **Minimal**: For laptop and workstation development, developers need the smallest Kubernetes solution possible. When running on Ubuntu, MicroK8s is a self-contained Kubernetes cluster that works with Azure AKS, Amazon EKS, and Google GKE.

- **Easy**: Lower administration and operations costs by using a single-package installation. All add-ons and dependencies are included.

- **Secured**: Updates are available for all security breaches and can be applied immediately or scheduled according to your maintenance cycle.

- **Up to date**: MicroK8s keeps up with upstream Kubernetes, releasing beta, RC, and final elements on the same day as upstream Kubernetes.

- **Complete**: A handpicked set of manifests for common Kubernetes capabilities and services is already included in MicroK8s:

  a. Automatic updates to the latest Kubernetes version

  b. Service mesh: Istio and Linkerd

  c. Serverless: Knative and OpenFaaS

  d. Monitoring: Fluentd, Prometheus, and Grafana, Metrics

e. Ingress, DNS, Dashboard, and clustering

f. GPU bindings for AI/ML

g. Cilium, Helm, and Kubeflow

Now that we know what MicroK8s is, let's see how easy it is to get started in the next section.

Quick installation

MicroK8s is deployed via **snaps**. Snaps are containerized (like Docker) software packages that are easy to create and install; they bundle their dependencies, and they work on all major Linux systems without modification. Snaps auto-update and are safe to run. Also, keep in mind that the MicroK8s snap will be updated frequently to keep up with Kubernetes releases.

In the next section, we will guide you through a minimal installation that'll work while we walk through the introduction.

Technical requirements

For the minimal installation, you require the following:

- You should have either a Linux distribution such as Ubuntu (20.04 LTS, 18.04 LTS, or 16.04 LTS) environment to execute the commands or any other operating system that supports `snapd`.

- 4 GB of memory and 20 GB of disk space are recommended.

Step 1 – Installation

In the following steps, we will be installing a MicroK8s cluster. We will install a limited set of components such as `api-server`, `controller-manager`, `scheduler`, `kubelet`, `cni`, and `kube-proxy`.

A MicroK8s snap can be installed using the following command:

```
sudo snap install microk8s --classic
```

The following command execution output confirms that MicroK8s has been installed successfully:

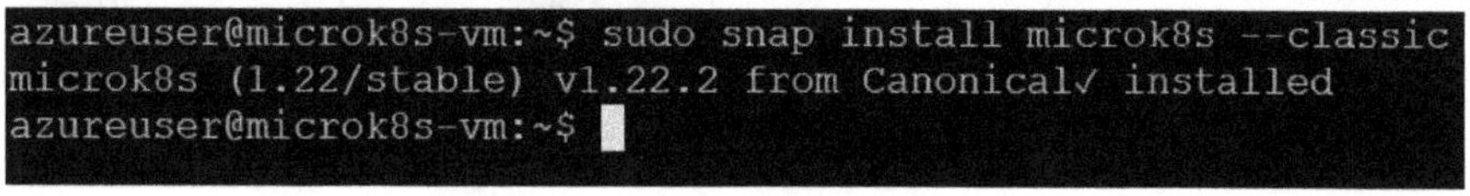

```
azureuser@microk8s-vm:~$ sudo snap install microk8s --classic
microk8s (1.22/stable) v1.22.2 from Canonical√ installed
azureuser@microk8s-vm:~$
```

Figure 2.1 – MicroK8s installation

Currently, I'm using an Ubuntu VM hosted on the cloud, but MicroK8s can also be installed on Windows, macOS, and Raspberry Pi on ARM hardware. For other platforms, please refer to the following link: `https//microk8s.io/docs/install-alternatives`.

> **Important Note**
>
> You can also specify a channel when installing MicroK8s. The specified channel consists of two parts – the *track* and the *risk* level. For example, to install MicroK8s v1.20 with the risk level set to `stable`, do the following:
>
> **`sudo snap install microk8s --classic --channel=1.20/stable`**
>
> When the MicroK8s team determines that a release (*edge* and *candidate*) is ready, your cluster is updated to the *stable* risk level, indicating that no bugs have been discovered by users running the same revision on riskier branches.

For accessing Kubernetes, MicroK8s includes its own version of `kubectl`. It can be used to perform commands that will monitor and control your Kubernetes cluster. MicroK8s adds a `microk8s.kubectl` command to avoid conflicting with an existing *kubectl* and overwriting any existing Kubernetes configuration file. If you just use MicroK8s, consider creating an alias with this command:

```
sudo snap alias microk8s.kubectl kubectl
```

The following command execution output confirms that an alias was added successfully:

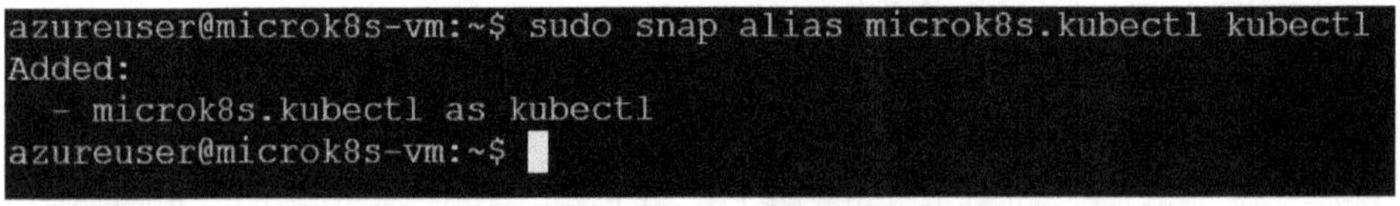

```
azureuser@microk8s-vm:~$ sudo snap alias microk8s.kubectl kubectl
Added:
   - microk8s.kubectl as kubectl
azureuser@microk8s-vm:~$
```

Figure 2.2 – kubectl – adding an alias

At this point, you have installed MicroK8s. In the next steps, we will be verifying whether the installation has succeeded or not.

Step 2 – Verify the installation

Next, check whether the newly deployed node is in the `Ready` state using the following command:

```
kubectl get nodes
```

The following is the command execution output:

```
azureuser@microk8s-vm:~$ kubectl get nodes
Insufficient permissions to access MicroK8s.
You can either try again with sudo or add the user azureuser to the 'microk8s' group:

    sudo usermod -a -G microk8s azureuser
    sudo chown -f -R azureuser ~/.kube

After this, reload the user groups either via a reboot or by running 'newgrp microk8s'.
azureuser@microk8s-vm:~$ 
```

Figure 2.3 – Verify the installation

If you get the error shown in *Figure 2.3*, it means MicroK8s doesn't have enough permissions. MicroK8s creates a group to make it easier to use commands that require administrative privileges. To acquire access to the .kube caching directory by adding the current user to the group, run the following two commands:

```
sudo usermod -a -G microk8s azureuser
sudo chown -f -R azureuser ~/.kube
```

The following is the output:

```
azureuser@microk8s-vm:~$ sudo usermod -a -G microk8s azureuser
azureuser@microk8s-vm:~$     sudo chown -f -R azureuser ~/.kube
azureuser@microk8s-vm:~$ 
```

Figure 2.4 – Adding users to the group

If you receive still an error, it means that MicroK8s is still starting the nodes in the background. Wait for a few minutes and try again. If the installation is successful, then you should be seeing the following output:

```
azureuser@microk8s-vm:~$ kubectl get nodes
NAME            STATUS    ROLES     AGE      VERSION
microk8s-vm     Ready     <none>    9m10s    v1.22.2-3+9ad9ee77396805
azureuser@microk8s-vm:~$ 
```

Figure 2.5 – Verify the installation

You can also use the `kubectl describe` command to get the details of the node as follows:

```
azureuser@microk8s-vm:~$ kubectl get nodes
NAME            STATUS   ROLES     AGE      VERSION
microk8s-vm    Ready    <none>    9m10s    v1.22.2-3+9ad9ee77396805
azureuser@microk8s-vm:~$ kubectl describe node microk8s-vm
Name:             microk8s-vm
Roles:            <none>
Labels:           beta.kubernetes.io/arch=amd64
                  beta.kubernetes.io/os=linux
                  kubernetes.io/arch=amd64
                  kubernetes.io/hostname=microk8s-vm
                  kubernetes.io/os=linux
                  microk8s.io/cluster=true
Annotations:      node.alpha.kubernetes.io/ttl: 0
                  projectcalico.org/IPv4Address: 10.1.0.4/24
                  projectcalico.org/IPv4VXLANTunnelAddr: 10.1.254.64
                  volumes.kubernetes.io/controller-managed-attach-detach: true
CreationTimestamp: Thu, 14 Oct 2021 08:40:48 +0000
Taints:           <none>
Unschedulable:    false
Lease:
  HolderIdentity:  microk8s-vm
```

Figure 2.6 – The describe command used on a node

At this point, you have a fully functional Kubernetes cluster. To summarize, we have installed MicroK8s and verified whether the installation was successful or not. In the next section, we are going to deploy a sample application on the MicroK8s cluster.

Deploying a sample application

We are going to deploy the `nginx` web server sample application. It is software that responds to client requests via **HTTP (Hypertext Transfer Protocol)**. The following command will deploy the `nginx` web application:

```
kubectl create deployment nginx --image=nginx
```

The following command execution output indicates that there is no error in the deployment, and in the next steps, we can verify whether the Pods have been created:

```
azureuser@microk8s-vm:~$ kubectl create deployment nginx --image=nginx
deployment.apps/nginx created
azureuser@microk8s-vm:~$ 
```

Figure 2.7 – Create the deployment

Check the `pods` status to verify whether the application has been deployed and is running:

```
kubectl get pods
```

The following command execution output indicates that Pods have been created and in the `Running` status:

```
azureuser@microk8s-vm:~$ kubectl get pods
NAME                      READY   STATUS    RESTARTS   AGE
nginx-6799fc88d8-s2z8n    1/1     Running   0          68s
azureuser@microk8s-vm:~$
```

Figure 2.8 – Check the status of the deployment

The `nginx` application has been deployed successfully, so it can be exposed with the following command:

```
kubectl expose deployment nginx \
--port 80 \
--target-port 80 \
--type ClusterIP \
--selector=run=nginx \
--name nginx
```

The following command execution output indicates that the `nginx` application is exposed successfully:

```
azureuser@microk8s-vm:~$ kubectl expose deployment nginx \
> --port 80 \
> --target-port 80 \
> --type ClusterIP \
> --selector=run=nginx \
> --name nginx
service/nginx exposed
azureuser@microk8s-vm:~$
```

Figure 2.9 – Expose the deployment

You should see a new service and the `ClusterIP` address assigned:

```
azureuser@microk8s-vm:~$ kubectl get svc
NAME         TYPE        CLUSTER-IP       EXTERNAL-IP   PORT(S)    AGE
kubernetes   ClusterIP   10.152.183.1     <none>        443/TCP    87m
nginx        ClusterIP   10.152.183.108   <none>        80/TCP     2m32s
azureuser@microk8s-vm:~$
```

Figure 2.10 – get svc and the ClusterIP address

Now that services are exposed externally, we can launch the web browser and point it to the external IP from our local machine to access the `nginx` application:

Welcome to nginx!

If you see this page, the nginx web server is successfully installed and working. Further configuration is required.

For online documentation and support please refer to nginx.org.
Commercial support is available at nginx.com.

Thank you for using nginx.

Figure 2.11 – The nginx landing page

Congratulations! You have now deployed the `nginx` application to a fully functional Kubernetes cluster by using MicroK8s. This will help you to understand how MicroK8s gets you up and running in under 60 seconds. In the next section, we are going to learn about add-ons and how to enable them.

Enabling add-ons

To provide a pure, lightweight version of Kubernetes, MicroK8s uses the bare minimum of components. With only a few keystrokes, **add-ons**, which are pre-packaged components that provide additional capabilities for your Kubernetes cluster, from simple DNS control to ML with Kubeflow, are accessible.

To begin, the DNS add-on should be enabled to promote communication between services. The storage add-on provides directory space on the host for programs that require storage.

These are easy to set up with the following command:

```
microk8s enable dns storage
```

The command execution output is as follows:

```
azureuser@microk8s-vm:~$ microk8s enable dns storage
Enabling DNS
Applying manifest
serviceaccount/coredns created
configmap/coredns created
Warning: spec.template.metadata.annotations[scheduler.alpha.kubernetes.
ctional in v1.16+; use the "priorityClassName" field instead
deployment.apps/coredns created
service/kube-dns created
clusterrole.rbac.authorization.k8s.io/coredns created
clusterrolebinding.rbac.authorization.k8s.io/coredns created
Restarting kubelet
DNS is enabled
Enabling default storage class
deployment.apps/hostpath-provisioner created
storageclass.storage.k8s.io/microk8s-hostpath created
serviceaccount/microk8s-hostpath created
clusterrole.rbac.authorization.k8s.io/microk8s-hostpath created
clusterrolebinding.rbac.authorization.k8s.io/microk8s-hostpath created
Storage will be available soon
azureuser@microk8s-vm:~$
```

Figure 2.12 – Enable DNS and storage add-ons

For the full list of available MicroK8s add-ons, please refer to the *Full list of add-ons* section.

Once the add-on is enabled, check whether all the components for the additional services can be started using the following command:

```
kubectl get all --all-namespaces
```

The following command execution output indicates (in the highlighted portions) that additional services have been started and are in Running status:

```
azureuser@microk8s-vm:~$ kubectl get all --all-namespaces
NAMESPACE     NAME                                           READY   STATUS    RESTARTS   AGE
default       pod/nginx-6799fc88d8-s2z8n                     1/1     Running   0          46m
kube-system   pod/calico-kube-controllers-6f896476f5-qnjsh   1/1     Running   0          129m
kube-system   pod/coredns-7f9c69c78c-87btk                   1/1     Running   0          33m
kube-system   pod/calico-node-msctq                          1/1     Running   0          129m
kube-system   pod/hostpath-provisioner-5c65fbdb4f-wznsp      1/1     Running   0          32m
```

Figure 2.13 – Verify that additional components have been started

Add-ons that have been enabled can be disabled at any time by utilizing the disable command:

```
microk8s disable dns
```

To check the list of available and installed addons, use the `status` command, as follows:

```
azureuser@microk8s-vm:~$ microk8s status
microk8s is running
high-availability: no
  datastore master nodes: 127.0.0.1:19001
  datastore standby nodes: none
addons:
  enabled:
    dns                    # CoreDNS
    ha-cluster             # Configure high availability on the current node
    registry               # Private image registry exposed on localhost:32000
    storage                # Storage class; allocates storage from host directory
  disabled:
    ambassador             # Ambassador API Gateway and Ingress
    cilium                 # SDN, fast with full network policy
    dashboard              # The Kubernetes dashboard
    fluentd                # Elasticsearch-Fluentd-Kibana logging and monitoring
    gpu                    # Automatic enablement of Nvidia CUDA
    helm                   # Helm 2 - the package manager for Kubernetes
    helm3                  # Helm 3 - Kubernetes package manager
    host-access            # Allow Pods connecting to Host services smoothly
    ingress                # Ingress controller for external access
    istio                  # Core Istio service mesh services
```

Figure 2.14 – Use the status command to check the list of available and installed add-ons

In case of errors, MicroK8s gives you troubleshooting tools to check out what has gone wrong. In the following sections, we can check how to use troubleshooting tools.

Full list of add-ons

The following table shows the current list of add-ons at the time of writing:

Add-on	Description
ambassador	Adds the Ambassador API and the ingress controller.
dashboard	Enables Kubernetes Dashboard.
dns	Deploys CoreDNS.
cilium	Deploys Cilium to support Kubernetes network policies using eBPF.
fluentd	Deploys the Elasticsearch-Fluentd-Kibana logging and monitoring solution.
gpu	Enables support for GPU accelerated workloads using the NVIDIA runtime.
helm	Deploys the Helm 2 package manager for Kubernetes.
helm3	Deploys the Helm 3 package manager for Kubernetes.

Add-on	Description
host-access	Provides a fixed IP for access to the host's services.
inaccel	Simplifies FPGA management and application life cycle with InAccel.
ingress	Enables the ingress controller for external access.
Istio	Adds the core Istio services.
jaeger	Deploys the Jaeger Operator with minimal configuration.
juju	Enables a Juju client to work with MicroK8s.
kata	Enables Kata containers support for secure container runtime with lightweight VMs.
keda	Deploys the **Kubernetes Event-Driven Autoscaling (KEDA)** operator.
knative	Enables Knative middleware to your cluster.
kubeflow	Adds Kubeflow using the Charmed Kubeflow operators.
linkerd	Deploys the Linkerd service mesh.
Metallb	Deploys the MetalLB load-balancer.
metrics-server	Adds the Kubernetes Metrics Server for API access to service metrics.
multus	Adds multus for multiple network capability.
openfaas	OpenFaaS, the popular serverless framework.
openebs	Adds OpenEBS storage capability.
portainer	Enables a Portainer solution for a container management dashboard.
prometheus	Deploys the Prometheus Operator.
rbac	Enables **Role-Based Access Control (RBAC)** for authorization.
registry	Deploys a private image registry and exposes it on localhost:32000. The storage add-on will be enabled as part of this add-on.
storage	Creates a default storage class that allocates storage from a host directory.
traefik	Adds the Traefik Kubernetes Ingress controller.

Table 2.1 – The complete list of MicroK8s add-ons

We have now understood what an add-on is, and we have enabled a few add-ons, such as dns and storage. We've also seen the entire list of add-ons. In the next section, we will look at how to start/stop a MicroK8s cluster.

Starting/stopping MicroK8s

MicroK8s will run indefinitely unless you instruct it to **stop**. With these simple commands, you can *stop and start* MicroK8s.

To stop a MicroK8s cluster, use the `microk8s stop` command:

```
azureuser@microk8s-vm:~$ microk8s stop
Stopped.
azureuser@microk8s-vm:~$
```

Figure 2.15 – Use the stop command to stop the MicroK8s cluster

To start the MicroK8s cluster, use the `microk8s start` command:

```
azureuser@microk8s-vm:~$ microk8s start
Started.
```

Figure 2.16 – Use the start command to start the MicroK8s cluster

MicroK8s will automatically restart after a reboot if you keep it running. If you don't want this to happen, simply run `microk8s stop` before turning off the computer.

We've seen how to start and stop a MicroK8s cluster. In the following section, we'll look at how to configure MicroK8s to work with local images.

Configuring MicroK8s to use local images

The **Kubernetes orchestration framework** uses container images to manage containerized applications. These images can be in a local filesystem or can be downloaded from a remote registry. The most popular container tool is **Docker**. The following diagram is an introduction to what Docker does:

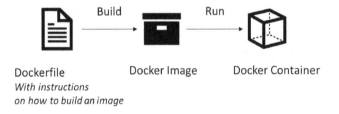

Figure 2.17 – What Docker does

Let's suppose we have a container image built and available in the local Docker image repository. For example, here, I have the `nginx1.21` image in the local Docker image repository:

```
azureuser@microk8s-vm:~$ docker images
REPOSITORY      TAG         IMAGE ID        CREATED        SIZE
nginx1.21       local       87a94228f133    13 days ago    133MB
nginx           latest      87a94228f133    13 days ago    133MB
hello-world     latest      feb5d9fea6a5    4 weeks ago    13.3kB
azureuser@microk8s-vm:~$ []
```

Figure 2.18 – Docker images from the local Docker repository

The `nginx1.21` local image is only recognized by Docker, and MicroK8s Kubernetes will not be aware of the image. This is because the MicroK8s Kubernetes cluster does not include your local Docker *daemon*. We can push the Docker image into the MicroK8s image cache by exporting it from the local Docker daemon using the following command:

```
azureuser@microk8s-vm:~$ docker save nginx1.21 > nginxlocal.tar
azureuser@microk8s-vm:~$ microk8s ctr image import nginxlocal.tar
unpacking docker.io/library/nginx1.21:local (sha256:f090a2b152845c78ed6b9eac73ffbc267abc0b06a8717ca798e89d9589e70511)...do
ne
azureuser@microk8s-vm:~$ []
```

Figure 2.19 – Push the Docker image

Now that we have imported the image to the MicroK8s image cache, we can confirm whether the image is in there by using the following command:

```
microk8s ctr images ls
```

The following command execution output shows that our `nginx1.21` image is available in the MicroK8s image cache:

```
azureuser@microk8s-vm:~$ microk8s ctr image ls | grep nginx1.21
docker.io/library/nginx1.21:local
ation/vnd.docker.distribution.manifest.v2+json          sha256:f090a2b152845c78ed6b9eac73ffbc267abc0b06a871
1 51.3 MiB  linux/amd64
-containerd.image=managed
azureuser@microk8s-vm:~$ []
```

Figure 2.20 – The list of containerd images

Now that we have the image, we can use the `microk8s kubectl apply -f <file>` command to deploy it to the MicroK8s Kubernetes.

Here, I have created the `nginx.local` file with the deployment instructions:

```
  GNU nano 4.8
apiVersion: apps/v1
kind: Deployment
metadata:
  name: nginx-deployment
  labels:
    app: nginx
spec:
  selector:
    matchLabels:
      app: nginx
  template:
    metadata:
      labels:
        app: nginx
    spec:
      containers:
      - name: nginx
        image: nginx1.21:local
        imagePullPolicy: Never
        ports:
        - containerPort: 80
```

Figure 2.21 – The nginx.local file with the deployment instructions

At this point, we are ready to deploy using the `kubectl apply` command:

```
azureuser@microk8s-vm:~$ microk8s kubectl apply -f nginx.local
deployment.apps/nginx-deployment created
azureuser@microk8s-vm:~$
```

Figure 2.22 – Create the deployment using the apply command

When the deployment is created, use the `kubectl get deployment` command to check the status of the deployments:

```
azureuser@microk8s-vm:~$ microk8s kubectl get deployment
NAME               READY   UP-TO-DATE   AVAILABLE   AGE
nginx              1/1     1            1           10d
nginx-deployment   1/1     1            1           6m10s
azureuser@microk8s-vm:~$
```

Figure 2.23 – Check the status of the deployment

The following fields are displayed:

- NAME indicates the names of the deployments in the namespace.
- READY indicates how many replicas of the application are available to the users of the applications.
- UP-TO-DATE indicates the number of replicas that have been updated to achieve the desired state.
- AVAILABLE indicates the number of replicas that are available to the users.
- AGE indicates the amount of time that the application has been operating.

Kubernetes will act as if there is an image in the Docker Hub registry at docker.io for which it already has a cached copy. This procedure can be repeated whenever the image is changed.

We have now seen how to work with locally built images without a registry. In the next section, we are going to look at how to use MicroK8s' built-in registry for the management of images.

Configuring MicroK8s to use its built-in registry

By minimizing the time spent on uploading and downloading Docker images, having a private Docker registry can help you to be more productive. The registry that comes with MicroK8s is hosted inside a Kubernetes cluster and is accessible as a NodePort service on the localhost's port 32000.

> **Important Note**
> You should be aware that this registry is not secured and will need additional steps to limit access from outside (in the case of production scenarios).

The first step is to enable the built-in registry using the following command:

```
azureuser@microk8s-vm:~$ microk8s enable registry:size=40Gi
Addon storage is already enabled.
Enabling the private registry
Applying registry manifest
namespace/container-registry created
persistentvolumeclaim/registry-claim created
deployment.apps/registry created
service/registry created
configmap/local-registry-hosting configured
The registry is enabled
The size of the persistent volume is 40Gi
azureuser@microk8s-vm:~$
```

Figure 2.24 – Enable the registry

As you can see, the registry add-on has been set up with a 40 Gi persistent volume claim for storing images. Please note that the storage add-on is also enabled along with the registry to enable storage claims.

Now that we have the registry set up, our next step is to tag the image and push it into the built-in registry:

```
azureuser@microk8s-vm:~$ docker tag 87a94228f133 localhost:32000/nginx1.21:registry
```

Figure 2.25 – Tag the Docker image

Push the tagged image into the built-in image, as shown here:

```
azureuser@microk8s-vm:~$ docker push localhost:32000/nginx1.21:registry
The push refers to repository [localhost:32000/nginx1.21]
9959a332cf6e: Pushed
f7e00b807643: Pushed
f8e880dfc4ef: Pushed
788e89a4d186: Pushed
43f4e41372e4: Pushed
e81bff2725db: Pushed
registry: digest: sha256:7250923ba35431100404462388756ef099331822c6172a050b12c7a38361ea46f size: 1570
azureuser@microk8s-vm:~$
```

Figure 2.26 – Push the tagged image

> **Important Note**
>
> Pushing to this insecure registry may fail in some Docker versions unless the daemon is specifically set to trust it.
>
> Add the registry endpoint (as shown in the following snippet) in /etc/docker/daemon. json and restart the Docker daemon:
>
> ```
> {
>    "insecure-registries" : ["localhost:32000"]
> }
> ```

Let's check whether the images are tagged by using the docker images command:

```
azureuser@microk8s-vm:~$ docker images
REPOSITORY                   TAG        IMAGE ID        CREATED        SIZE
nginx1.21                    local      87a94228f133    13 days ago    133MB
nginx                        latest     87a94228f133    13 days ago    133MB
localhost:32000/nginx1.21    registry   87a94228f133    13 days ago    133MB
hello-world                  latest     feb5d9fea6a5    4 weeks ago    13.3kB
azureuser@microk8s-vm:~$
```

Figure 2.27 – Check whether the images are tagged

Now that we have the image, we can use the `kubectl apply -f <file>` command to deploy it to the MicroK8s Kubernetes.

Here, I have created the `nginx.builtin` file with the deployment instructions:

```
  GNU nano 4.8
apiVersion: apps/v1
kind: Deployment
metadata:
  name: nginx-builtin-registry-deployment
  labels:
    app: nginx
spec:
  selector:
    matchLabels:
      app: nginx
  template:
    metadata:
      labels:
        app: nginx
    spec:
      containers:
      - name: nginx
        image: localhost:32000/nginx1.21:registry
        ports:
        - containerPort: 80
```

Figure 2.28 – The nginx.builtin file with the deployment instructions

At this point, we are ready to deploy using the `kubectl apply` command:

```
azureuser@microk8s-vm:~$ microk8s kubectl apply -f nginx.builtin
deployment.apps/nginx-builtin-registry-deployment created
azureuser@microk8s-vm:~$
```

Figure 2.29 – Create the deployment using the apply command

When the deployment is created, use the `kubectl get deployment` command to check the status of the deployments:

```
azureuser@microk8s-vm:~$ microk8s kubectl get deployment
NAME                                    READY   UP-TO-DATE   AVAILABLE   AGE
nginx                                   1/1     1            1           11d
nginx-deployment                        1/1     1            1           167m
nginx-builtin-registry-deployment       1/1     1            1           68s
azureuser@microk8s-vm:~$
```

Figure 2.30 – Check the status of the deployments using the kubectl get deployment command

Kubernetes will pull the image from the built-in registry. If there is a change to the image, *building* and *pushing* the images to the built-in registry can be repeated so that updates are propagated. If you've dedicated machines that host registries in a cluster, you'll need to alter the configuration files to point to the node's IP address.

In the next section, we will go through the steps on how to configure MicroK8s to pull the images from any of the public or private registries.

Configuring MicroK8s to use private/public registries

MicroK8s can pull the images from *private* or *public* registries as well, but before being able to pull container images, MicroK8s Kubernetes must be made aware of the registry endpoints.

Let's assume that a **private registry** is set up at some IP address, such as 10.131.231.155. The images we build need to be tagged with the IP address:Port/image:tag registry endpoint syntax, as follows:

```
docker build . -t 10.131.231.155:32000/nginx1.21:registry
```

> **Important Note**
>
> Pushing to this insecure registry may fail in some Docker versions unless the daemon is specifically set to trust it. Add the registry endpoint (as shown in the following snippet) to /etc/docker/daemon.json and restart the Docker daemon:
>
> ```
> {
>    "insecure-registries" : ["10.131.231.155:32000"]
> }
> ```

To push it to one of the **public registries** such as hub.docker.com, use the docker login command to log in and push the image tagged with docker-hub-username/image-name:tag:

```
docker tag 87a94228f133 10.131.231.155:32000/nginx1.21:registry
```

Once the image is tagged, push the tagged image into a private or public registry:

```
docker push 10.131.231.155:32000/nginx1.21:registry
```

Now that we have the image, we can use the kubectl apply -f <file> command to deploy like earlier. Here, I have created a file with the deployment instructions:

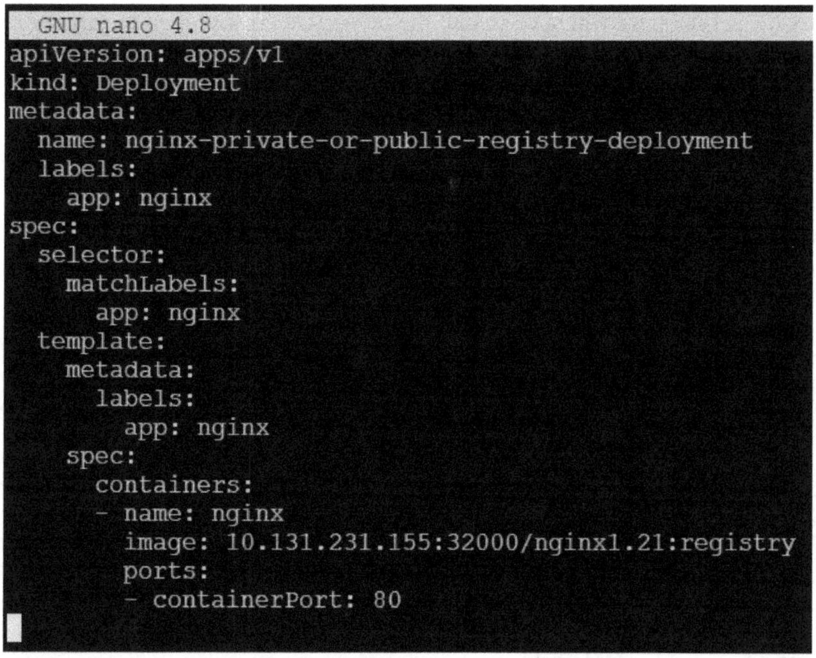

```
  GNU nano 4.8
apiVersion: apps/v1
kind: Deployment
metadata:
  name: nginx-private-or-public-registry-deployment
  labels:
    app: nginx
spec:
  selector:
    matchLabels:
      app: nginx
  template:
    metadata:
      labels:
        app: nginx
    spec:
      containers:
      - name: nginx
        image: 10.131.231.155:32000/nginx1.21:registry
        ports:
        - containerPort: 80
```

Figure 2.31 – The deployment file with instructions

Once the deployment is created, use the kubectl get deployment command to check the status of the deployments.

> **Important Note**
>
> In the *production* scenarios, a **private secure registry** needs to be used, which is more secure and limits access to specific users/applications. The recommended way is to create a secret from the Docker login credentials and use this secret to access the secure registry.

To recap, we looked at how to set up MicroK8s using either local images or ones fetched from public or private registries. In the next section, we will look at how to configure the various services or components of MicroK8s.

Configuring MicroK8s services

MicroK8s is made up of various services or components that are managed by a number of system daemons. Configuration of these services is read from files stored in the $SNAP_DATA directory, which normally points to /var/snap/microk8s/current.

To reconfigure the services, we will need to edit the respective file and then restart the respective daemon. The following table shows *system daemon services* that will be run by MicroK8s:

MicroK8S service	Description	Path
`snap.microk8s.daemon-apiserver`	A daemon for the Kubernetes API server that verifies and configures data for API objects such as Pods, services, and replication controllers	`${SNAP_DATA}/args/kube-apiserver`
`snap.microk8s.daemon-containerd`	A daemon for the `containerd` container runtime to manage images and execute containers	`${SNAP_DATA}/args/containerd & ${SNAP_DATA}/args/containerd-template.toml`
`snap.microk8s.daemon-controller-manager`	A daemon for the Kubernetes controller manager that embeds the Kubernetes core control loops	`${SNAP_DATA}/args/kube-controller-manager`
`snap.microk8s.daemon-etcd`	A daemon for the `etcd` key-value cluster datastore for Kubernetes components	`${SNAP_DATA}/args/etcd`
`snap.microk8s.daemon-flanneld`	A daemon for the Flannel **Container Network Interface** (**CNI**) that assigns each host a subnet for use with container runtimes	`${SNAP_DATA}/args/flanneld`
`snap.microk8s.daemon-kubelet`	A daemon for the `kubelet` node agent that runs on each node	`${SNAP_DATA}/args/kubelet`

MicroK8S service	Description	Path
`snap.microk8s.daemon-kubelite`	A daemon for `kubelite` that runs as subprocesses of the scheduler, controller, proxy, kubelet, and `apiserver` services	• Scheduler: `${SNAP_DATA}/args/kube-scheduler` • Controller: `${SNAP_DATA}/args/kube-controller-manager` • Proxy: `${SNAP_DATA}/args/kube-proxy` • kubelet: `${SNAP_DATA}/args/kubelet` • apiserver: `${SNAP_DATA}/args/kube-apiserver`
`snap.microk8s.daemon-proxy`	A daemon for the Kubernetes network proxy that runs on each node	`${SNAP_DATA}/args/kube-proxy`
`snap.microk8s.daemon-scheduler`	A daemon for the Kubernetes scheduler that considers individual and collective resource requirements, quality of service requirements, hardware/software/policy constraints, affinity and anti-affinity specifications, data locality, inter-workload interference, deadlines, and so on	`${SNAP_DATA}/args/kube-scheduler`

Table 2.2 – A list of the MicroK8s system daemon services

In the next section, we will look at how to troubleshoot issues at the application and cluster levels.

Troubleshooting application and cluster issues

It's critical to know that things can go wrong; there might be issues with the Kubernetes components themselves, or a problem with the MicroK8s component. In this section, we will cover some of the common issues and tools to assist you in determining what went wrong.

The application level

This section assists users in debugging Kubernetes-deployed applications that aren't functioning as intended.

Examining a **Pod** is the first step in troubleshooting it. With the following command, you can check the current state of the Pod and a historical list of the events:

```
kubectl describe pods ${POD_NAME}
```

In the following command execution output, you can see that the `kubectl describe pod` command fetches the details of the container(s) and the Pod's configuration information (labels, resource needs, and so on), as well as the container(s) and Pod's status information (state, readiness, restart count, events, and so on):

```
azureuser@microk8s-vm:~$ kubectl describe pods nginx-deployment-5476f8c6df-ttkzz
Name:           nginx-deployment-5476f8c6df-ttkzz
Namespace:      default
Priority:       0
Node:           microk8s-vm/10.1.0.4
Start Time:     Mon, 25 Oct 2021 08:00:08 +0000
Labels:         app=nginx
                pod-template-hash=5476f8c6df
Annotations:    cni.projectcalico.org/podIP: 10.1.254.89/32
                cni.projectcalico.org/podIPs: 10.1.254.89/32
Status:         Running
IP:             10.1.254.89
IPs:
  IP:           10.1.254.89
Controlled By:  ReplicaSet/nginx-deployment-5476f8c6df
Containers:
  nginx:
    Container ID:   containerd://597bc9b01fdffd9b6dce344f62f9c746b3c71ac53929f2cfe267181af
    Image:          nginx1.21:local
    Image ID:       docker.io/library/nginx@sha256:644a70516a26004c97d0d85c7fe1d0c3a67ea8a
    Port:           80/TCP
    Host Port:      0/TCP
    State:          Running
      Started:      Thu, 28 Oct 2021 08:09:24 +0000
    Last State:     Terminated
      Reason:       Unknown
      Exit Code:    255
```

Figure 2.32 – Troubleshooting with the kubectl describe pods command

The following table summarizes all the possible issues and solutions based on the Pod state from the previously described command:

Pod state	What is the issue?	Possible solutions
PENDING	It cannot be scheduled on to a node.	This could be due to insufficient resources of one type or another that prevent scheduling. Look at the output of `kubectl describe <pod name>` to investigate why it cannot be scheduled.
WAITING	The Pod has been scheduled, but it can't be run on that node.	This could be due to the name of the image being misspelled, the image being in a repository, and so on. Look at the output of `kubectl describe <pod name>` to investigate why the Pod is not able to run.
RUNNING or crashed		Examine the logs of the Pod to troubleshoot the applications that are running or crashing. Look at the output of `kubectl logs <pod name>` to investigate why the applications are crashing, or debug the container using the `kubectl exec <pod name> -c <container name>` command.

Table 2.3 – Examining the Pod status for possible issues

We've looked at how to infer application issues from the Pod status information; in the following part, we'll look at it at the cluster level.

The cluster level

To get detailed information about the overall health of a MicroK8s cluster, you can run the following command:

```
microk8s kubectl cluster-info dump
```

By default, this command dumps all the information about the cluster to the output for debugging and diagnosing cluster problems. It also dumps the logs of all the Pods in the cluster, which are split into directories by namespace and Pod name.

Kubernetes also generates events whenever any of the resources it manages undergoes a change. The entity that initiated the event, the type of event (`Normal`, `Warning`, `Error`, and so on), and the cause are generally included in these events. This data is typically stored in `etcd` and made available when you run `kubectl events` commands.

These events provide insight into what occurred behind the scenes when a specific entity entered a particular state:

```
microk8s kubectl get events
```

MicroK8s also has a built-in inspection tool to generate a comprehensive report on the status of MicroK8s subsystems and the machine they run on. By running the tool, we can verify whether the system is working or not, and it also gathers all the important data for bug reporting:

```
sudo microk8s inspect
```

Administrator privileges are required to run this tool and gather the data:

```
azureuser@microk8s-vm:~/snap/microk8s$ sudo microk8s inspect
Inspecting Certificates
Inspecting services
  Service snap.microk8s.daemon-cluster-agent is running
  Service snap.microk8s.daemon-containerd is running
  Service snap.microk8s.daemon-apiserver-kicker is running
  Service snap.microk8s.daemon-kubelite is running
  Copy service arguments to the final report tarball
Inspecting AppArmor configuration
Gathering system information
  Copy processes list to the final report tarball
  Copy snap list to the final report tarball
  Copy VM name (or none) to the final report tarball
  Copy disk usage information to the final report tarball
  Copy memory usage information to the final report tarball
  Copy server uptime to the final report tarball
  Copy current linux distribution to the final report tarball
  Copy openSSL information to the final report tarball
  Copy network configuration to the final report tarball
Inspecting kubernetes cluster
  Inspect kubernetes cluster
Inspecting juju
  Inspect Juju
```

Figure 2.33 – The MicroK8s inspection tool

We recognized that things may go wrong; there can be issues with the application, Kubernetes components, or the MicroK8s component. We've looked at how to diagnose problems and picked up a few tools to help us with our debugging.

If you are unable to resolve your problem and feel it is due to a bug in MicroK8s, please submit an issue to the project repository at the following link: https://github.com/ubuntu/microk8s/issues/.

Summary

To summarize, we learned how to install MicroK8s, check the progress of the installation, and monitor and control a Kubernetes cluster in this chapter. We learned how to install a sample application and use some of the add-ons as well.

We also learned how to use MicroK8s with local container images as well as images retrieved from public and private registries. Furthermore, we investigated the inspection tool that creates a complete report on MicroK8s and the system it runs on, as well as walked through common issues to assist in fixing the most frequently encountered problems.

The key concepts of *edge computing* will be introduced in the next chapter. We'll also look at some of the things to keep in mind when you develop your edge architecture.

Part 2: Kubernetes as the Preferred Platform for IoT and Edge Computing

Data volumes continue to grow, particularly in industries such as manufacturing, oil and gas, energy, and transportation that are undergoing rapid digital transformation. There is a need to manage this data explosion at the edge and the many associated challenges, including the complexity of systems, data privacy, latency issues, low bandwidth connectivity, and increasing costs for storing and processing data, either in the cloud or data centers. In this part, we will look at how Kubernetes, the edge, and the cloud can collaborate to drive intelligent business decisions.

This part of the book comprises the following chapters:

- *Chapter 3, Essentials of IoT and Edge Computing*
- *Chapter 4, Handling the Kubernetes Platform for IoT and Edge Computing*

3
Essentials of IoT and Edge Computing

Data volumes continue to grow, particularly in industries such as manufacturing, oil and gas, energy, and transportation that are undergoing rapid digital transformation. This data explosion at the edge must be managed. There are numerous challenges involved with this, including system complexity, data privacy, latency issues, poor bandwidth connectivity, and rising costs for storing and processing data.

Edge computing reduces the amount of long-distance connectivity between a device and server by getting compute as close to the data source as possible, thereby improving the way the data is handled, processed, and delivered. When it comes to large industries such as manufacturing, oil and gas, energy, or transportation, industrial edge computing is being deployed to analyze and manage all data at the asset end in real time for real-time analytics or to use the aggregated data for further processing in the cloud.

The edge has three main components, as outlined here:

- **Connectivity**: The ability to connect to any industrial system and collect and normalize data for immediate use

- **Intelligence**: Focusing on data processing and analytics functions at the edge to act and derive insights at the data source

- **Orchestration**: The ability to create, deploy, manage, and update edge applications

Edge computing allows businesses to operate closer to the origin of the data more efficiently and with reduced latency. For rapid business decisions that improve quality and processes, the edge unlocks real-time analytics such as inventory usage, asset uptime and downtime, capacity utilization, and more.

The edge enables predictive/prescriptive maintenance, condition-based monitoring, overall equipment effectiveness, vision systems, quality improvements, and more. Edge data can also enable more advanced use cases such as **artificial intelligence (AI)** and **machine learning (ML)**. The intelligent edge is driving significant operations and process improvements.

Businesses have begun to adopt modern edge platforms to drive these initiatives with a unified solution that enables the three facets of the edge—edge computing, edge analytics, and edge intelligence. A recent report from *Gartner* states that deployed **Internet of Things (IoT)** endpoints in the manufacturing and natural resource industries are expected to reach 1.9 billion units by 2028.

In this and the following chapter, we are going to look at how Kubernetes, the edge, and the cloud can collaborate to drive intelligent business decisions.

In this chapter, we're going to cover the following main topics:

- What is IoT?
- What is edge computing?
- How are IoT and the edge related?
- Benefits of edge computing
- What does it take to enable edge computing, edge analytics, and edge intelligence?

What is IoT?

According to *Gartner* (`https://www.gartner.com/en/information-technology/glossary/internet-of-things`), IoT is a network of physical objects that contain embedded technology to communicate and sense or interact with their internal states or the external environment. In simple terms, IoT refers to the process of connecting all the world's devices to the internet.

IoT devices can range from everyday objects such as lightbulbs to healthcare assets such as medical gadgets, wearables, smart devices, and even traffic lights in smart cities.

Here are a few examples of IoT use cases:

- **Virtual reality (VR)** and **augmented reality (AR)**
- Smart cities
- Industrial IoT
- Self-driving cars
- Smart thermostats
- Smart homes
- Smartwatches

A typical IoT system sends, receives, and analyzes data continuously in a feedback loop. Humans or AI and ML can do analysis in near real time or over a longer period of time. *Gartner* estimated that there would be 20 billion internet-connected objects by 2020. This is the case, even though these are not general-purpose gadgets, but rather products such as jet engines, connected cars, or even coffee machines.

Data is collected and processed using an IoT edge device. An IoT edge device has the storage and computational capability to make low-latency decisions and analyze data in milliseconds; however, not all edge devices are IoT devices. In the sections that follow, we'll go through all this in greater depth.

Businesses can employ IoT solutions to evaluate data generated by IoT edge devices and use this information to improve business decisions and develop new business applications. In the next section, we'll look at a typical IoT solution design that allows you to integrate and work together with data from a variety of systems, devices, apps, and human interactions.

Key elements of an IoT solution

There is a gap between IoT edge devices and corporate applications due to a lack of interface and communication standards. An IoT platform acts as a middleware that connects the two endpoints, bridging the gap. Modern IoT platforms, on the other hand, go a step further than this by including functionality in both the hardware and application layers. As a result, the functionality set of an IoT platform may include capabilities for edge data processing or complex data analysis techniques.

Devices that are connected to the internet—such as smartphones, digital watches, and electrical appliances—begin the process by securely communicating with the IoT platform. The platforms collect and analyze data from a variety of devices and then send the most valuable data to devices via applications.

IoT edge devices that support important communication technologies, including Wi-Fi, **NarrowBand-IoT (NB-IoT)**, and Sigfox, as well as protocols such as **Message Queuing Telemetry Transport (MQTT)**, **Constrained Application Protocol (CoAP)**, and **HyperText Transfer Protocol Secure (HTTPS)**, can connect directly to the IoT platform. Because this isn't always the case, an abstraction layer is required, which provides the technologies and procedures required to translate between multiple protocols and communication technologies, thereby mediating data flow.

A typical IoT platform's architecture is represented in the following diagram—it allows you to connect and collaborate with data from many systems, devices, apps, and human interactions:

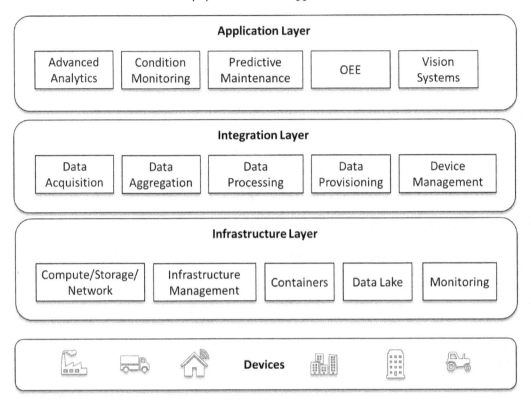

Figure 3.1 – IoT platform architecture

Here is a brief description of each layer of the architecture:

- The **infrastructure** layer contains components that allow the platform to function as a whole. Here, you'll find various compute/storage/network solutions, container management options, data lakes, internal platform messaging, as well as monitoring and storage solutions.

- The **integration** layer is primarily in charge of receiving data from connected edge devices, analyzing it, distributing it to business applications, and managing devices.

- The **security** layer oversees assuring the data's validity, security, and privacy by enforcing data safeguard and authority principles, as well as implementing remedial controls and actions.

- The **application** layer contains apps that use the integration layer's fundamental IoT features to meet business objectives. This is where applications such as complex data analytics, condition monitoring, retrofitting/configuration, and predictive maintenance applications would be housed.

Now that we've looked at the basic architecture of an IoT platform, let's look at some of the challenges of IoT. In recent years, the volume of sensor-generated data from gadgets has exploded, and this trend is projected to continue. Because data loses its value quickly after creation, frequently in milliseconds, the speed with which businesses can turn data into insights and subsequently into action is considered mission-crucial. As a result, a company's agility can be enhanced by minimizing the time between data generation and decision or action as far as possible.

However, because data transmission speed is intrinsically limited by the speed of light, latency can only be reduced or eliminated totally by reducing the distance that data must travel. Because data travels hundreds or even thousands of miles in a cloud-only world, edge computing can be used in circumstances where latency is critical. Let's look at how edge computing addresses this issue in the next section.

What is edge computing?

Edge computing is a distributed computing concept in which intelligence is built into edge devices, also known as edge nodes, allowing data to be processed and analyzed in real time close to the data collection point. When adopting edge computing, data does not need to be transferred to the cloud or to a centralized data processing system.

According to *Gartner*'s estimate, up to 55 percent of IoT data could be processed at the origin in the near future, either on the device itself or through edge computing. Indeed, scalability will likely play a key role in this change, as increased data demands will necessarily focus on latency, and lowering latency might significantly improve reaction time, saving time and money.

A typical edge computing architecture is shown in the following diagram:

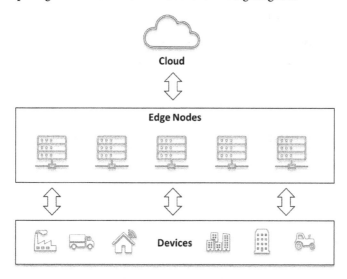

Figure 3.2 – Edge computing architecture

Because data is handled close to the data aggregation source, edge computing eliminates the need to move data to the cloud or an on-premises data center for processing and analysis. Both the network and the servers are less stressed as a result of this strategy. Edge computing is ideal for IoT, particularly **industrial IoT (IIoT)**, because of its ability to process data in real time and its faster response time.

Edge computing technology provides far more breakthroughs, such as AI and ML, in addition to faster digital transformation for industrial and manufacturing organizations.

Let's look at some specific use cases of edge computing here:

- **Visual inferencing**: High-resolution cameras are frequently installed on edge devices. They can consume video streams from these cameras and do ML on the information right on the edge machine. That inferencing can just detect the persons in the view, or it can execute more complicated inferences. Additionally, some edge applications can detect whether someone is wearing a correctly fitted mask, has a high body temperature, or is keeping a suitable distance between them.

- **Anomaly detection**: In many industrial plants and factories, edge devices play a significant role. They can see and hear things that humans are unable to discern, and they can infer difficulties as well as (or sometimes even better than) the most expert human operators in a variety of situations. Edge computers, for example, can precisely monitor electric motor power use and respond quickly without the need for human participation. Edge computers can also listen for unusual sounds coming from sophisticated machinery and inform human operators if something is wrong.

- **Environment monitoring**: Edge devices can detect dangerous conditions such as particulate pollution or poisonous gases and act quickly (within milliseconds) to remedy or minimize the situation while also contacting the proper authorities.

- **Multi-access edge computing** (MEC): Since they are geographically near to users, computers in telecom MEC facilities can provide services that were previously only available in huge corporate data centers or public clouds, but with substantially lower latencies. Because the edge devices in these MECs are similar to the server class machines found in major data centers, they can handle any data center workload.

Now that we have seen some of the use cases of edge computing, let's look at how IoT and the edge are related in the next section.

How are IoT and the edge related?

IoT benefits from processing power being closer to the physical device or data source. For data collected by IoT sensors and devices to be used to react faster or eliminate issues, it must be analyzed at the edge rather than traveling back to a central site.

By providing a local source of processing and storage for the data and computing demands of IoT devices, edge computing reduces communication latency between IoT devices and the central **information technology (IT)** networks to which they are connected.

IoT would rely on cloud or data center connection and computation services without edge computing. Sending data back and forth between an IoT device and the cloud could slow response times and reduce operational efficiency.

Edge computing also addresses network capacity requirements for transferring large amounts of data across sluggish cellular or satellite connections, as well as the ability for systems to work offline if a network connection is lost.

The huge amounts of data created by linked IoT devices can be taken advantage of by using edge computing. Data may be processed locally and used to make quick decisions by deploying analytics algorithms and ML models to the edge. Data can also be aggregated locally before being sent to a central location for processing or long-term storage.

Under the cloud computing model, compute resources and services are usually centralized at large data centers. The clouds frequently provide the network infrastructure required to connect IoT devices to the internet. Edge devices require network access to provide back-and-forth communication between the device and a central database. The clouds are often used to provide network connectivity. The transmission of data from an edge device to a data center via the cloud, or an edge device transmitting a record of its decisions back to the data center for data storage, processing, or big data analysis, are examples of the cloud's communication capabilities.

Many IoT and edge use cases stem from the need to process data locally in real-time situations, where sending the data to a data center for processing would result in unacceptable delay. Here are some examples of this:

- *IoT sensors create a continual stream of data on the manufacturing floor* that may be utilized to prevent breakdowns and optimize operations. A modern plant with 2,000 pieces of equipment can generate around 2,200 **terabytes (TB)** of data every month, according to one estimate. Instead of sending this wealth of data to a remote data center, it's faster and less expensive to process it closer to the equipment. However, connecting the equipment through a centralized data platform is still desirable. Equipment, for example, can receive standardized software upgrades and transmit filtered data that can aid in the improvement of operations in other plant locations.

- *Another common example of edge computing is connected automobiles.* Computers are installed on buses and railways to track passenger flow and service delivery. With the technology onboard their trucks, delivery drivers can determine the most effective routes. When employing an edge computing strategy, each vehicle runs on the same standardized platform as the rest of the fleet, improving service reliability and maintaining data security across the board.

- **Network functions virtualization** (**NFV**), which uses virtual computers running on conventional hardware at the network edge, is becoming increasingly popular among telecom operators. These **virtual machines** (**VMs**) can take the place of costly proprietary hardware. With an edge computing strategy, suppliers can keep the software running reliably and with uniform security standards in tens of thousands of remote sites. In a mobile network, applications running close to the end user reduce latency and allow providers to offer new services.

The following diagram depicts how sensors and gadgets on a factory floor interact with the cloud and the edge:

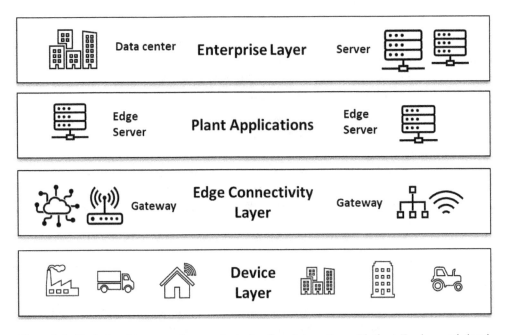

Figure 3.3 – Typical architecture of a manufacturing floor interacting with the IoT, edge, and cloud

Here is a brief description of each layer:

- **Device layer**: This layer represents bits and pieces of major equipment or the actual equipment itself and is tied to local operational technologies and IoT capabilities for rapid interactions. This layer uses cloud-trained ML models to perform ML scoring or inferencing. There's also a lot of raw device data stored here.

- **Edge connectivity and plant applications layers**: While the plant applications layer provides visibility and control of all bits and pieces of equipment or the equipment itself in a plant, the edge connection layer allows the equipment and plant applications to communicate with one another.

- **Cloud-hosted enterprise layer**: A portfolio view is provided by the cloud-hosted enterprise layer, which primarily gives a view and control across multiple plants. Business analysis and ML algorithms are part of this layer to predict and provide actionable intelligence using trained ML models. These models could be retrained in this layer to use data from across all plants' equipment portfolios and are then sent back to the edge and eventually to the IoT software at each piece of equipment, making operations smarter with the data.

Benefits of edge computing

To summarize, data can be analyzed, processed, and transported at the network's edge via edge computing. In other words, the data is analyzed in real time, close to where it is kept, with no latency. Data from IoT devices can be analyzed at the network's edge before being sent to a data center or cloud.

Some of the advantages of edge computing are listed here:

- High speed, low latency, and great dependability, allowing for faster data processing and content delivery

- Improved security by dispersing processing, storage, and applications across a diverse set of devices and data centers, making it more difficult for a single outage to bring the network down

- Provides a much more cost-effective path to scalability and versatility, allowing businesses to enhance their computing capacity by combining IoT devices with edge data centers

While there are many benefits to edge computing, it can also provide operational and architectural challenges. Edge processing, which is highly decentralized and generally comprises far-flung and/or difficult-to-access locations, includes sensors/actuators and gateways in offices, plants, campuses, on pipelines, and in other remote field sites. In any given organization, there are thousands of devices and hundreds of gateways. These edge nodes are equipped with firmware, operating systems, virtualization and containers, and applications, some of which are provided by manufacturers and others by solution providers. These edge nodes must be properly managed and maintained by their owner/manager, who necessitates a high level of automation (such as for backups, patching, updates, and monitoring).

Some operational and design challenges that need to be taken care of are outlined here:

- **Updating of edge nodes**: Edge processing involves the execution and replication of data center operations such as provisioning, updating, change management, and monitoring across all edge nodes and clusters, as well as additional services, such as device management and updating ML models.

- **Policies and practices**: Edge deployments, which are scattered over multiple sites and are significantly more dynamic than traditional data centers, are typically unsuited for traditional data center regulations and processes. Taking on the operational administration of such a system is difficult.

- **Cost**: While the cloud offers on-demand scalability and is simple to alter, automate, and maintain, implementing similar functionality at the edge can be costly and time-consuming. Extending an existing edge deployment to accommodate more devices and edge nodes can require a significant investment in more hardware and software, as well as a significant amount of time and work.

- **Cybersecurity**: When the cloud and data center are extended to the perimeter with numerous nodes and devices, the surface area for cyberattacks increases considerably. Insecure endpoints, such as devices and edge nodes, can be used to gain access to a company's valuable assets as well as for other purposes such as **distributed denial-of-service (DDoS)** assaults. It's a challenging and vital duty to keep all assets on the edge in a physically secure state.

The edge is just a necessary or mandatory feature of many IoT use cases if there is currently an operational technology. Adding a cloud-hosted component to an IoT system, even if it's simply a gateway, demands some kind of edge computing presence. The use of certain edge-processing capacity, for example, is required to add smart features to current system architecture and provide a cloud-based portfolio view. We'll look at what it takes to enable edge computing in the next section.

What does it take to enable edge computing, edge analytics, and edge intelligence?

To meet the needs of modern manufacturing, automotive, or telecommunications industries, it is important to acknowledge that businesses are under tremendous pressure to drive innovation and efficiency on their plant floors. This necessitates a comprehensive methodology that combines new/existing IT products with emerging methodologies, along with the goal of avoiding the time-consuming and error-prone manual configuration of several devices and applications at scale.

The following considerations must be noted while building an edge architecture:

- **Autonomy and resiliency**: Because the solution necessitates autonomy, connection interruptions cannot be permitted.

- **Resource constraints**: Low compute capability and small footprint of the devices.

- **Security challenges**: Data privacy, physical device security, and network security of the connected devices.

- **Manageability**: Manage application software across thousands of devices from many different suppliers.

- **Reliability**: Consistency in the building, deployment, and maintenance of applications.

- **Automation**: With high levels of automation, provision for automated mechanisms to deploy and maintain multiple distributed applications over any number of physical or virtual computers.

Given the difficulties of putting up an edge-based architecture, it is best to decide whether edge computing is necessary. A cloud-only solution may be the best option. The next step is to determine which capabilities are necessary at the edge, and then to choose the optimum deployment plan, given that edge processing can occur on devices, gateways, edge servers, possibly over multiple tiers, or in micro data centers.

Summary

In conclusion, IoT and the data it generates are transforming the world and how we interact with it. The cloud is at the heart of most of the connected-consumer IoT world, thanks to its numerous advantages. An IoT solution, on the other hand, will almost always involve both the edge and the cloud. Bringing computing to the edge helps organizations make better, faster decisions and become more agile by reducing latency, increasing scalability, and improving information access.

When deciding on the correct balance of edge and cloud capability for an IoT solution, it's important to remember that edge computing comes in a variety of flavors, each with its own set of advantages and drawbacks. Significant operational difficulties and costs can arise quickly, so businesses should examine a wide range of issues when planning and implementing any IoT system.

In the next chapter, we will look at how Kubernetes for the edge offers a compelling value proposition and different architectural approaches that demonstrate how Kubernetes can be used for edge workloads, as well as support for the architecture that meets enterprise applications' requirements—low latency, resource-constrained, data privacy, bandwidth scalability, and so on.

Handling the Kubernetes Platform for IoT and Edge Computing

Kubernetes has achieved significant adoption in data center and cloud environments since its launch in 2014. Kubernetes has progressed from orchestrating lightweight application containers to managing and scheduling a diverse set of IT workloads, ranging from virtualized network operations to AI/ML and GPU hardware resources.

Kubernetes is quickly gaining popularity as the most popular control plane for scheduling and managing work in distributed systems. These activities could involve deploying virtual machines on real hosts, pushing containers to edge nodes, and even expanding the control plane to incorporate additional schedulers, such as serverless environments. Its extensibility makes it a universal scheduler and the most preferred management platform. In this chapter, we are going to explore various deployment approaches to how Kubernetes, the edge, and the cloud can collaborate to drive intelligent business decisions.

Reiterating the points we discussed in the last chapter, the following considerations must be noted while building the edge architecture:

- **Autonomy and resiliency**: Because the solution necessitates autonomy, connection interruptions cannot be permitted.

- **Resource constraints**: Low compute capability and small device footprints.

- **Security challenges**: Data privacy, physical device security, and the network security of the connected devices.

- **Manageability**: Manage application software across thousands of devices from many different suppliers.

- **Reliability**: Consistency in the building, deployment, and maintenance of applications.
- **Automation**: With high levels of automation, provision for automated mechanisms to deploy and maintain multiple distributed applications over any number of physical or virtual computers.

Let's look at four architectural approaches that meet these criteria. In this chapter, we're going to cover the following main topics:

- Deployment approaches for edge computing
- Propositions that Kubernetes offers

Deployment approaches for edge computing

The following approaches demonstrate how Kubernetes can be used for edge workloads, as well as support for the architecture that meets enterprise applications' requirements – low latency, resource constraints, data privacy, bandwidth scalability, and others.

Deployment of the entire Kubernetes cluster at the edge

The complete Kubernetes cluster is deployed among edge nodes in this approach. This solution is better suited to situations where the edge node has limited capacity and does not want to consume additional resources for control planes and nodes.

The simplest production-grade upstream **K8s** is **MicroK8s**, a **Cloud Native Computing Foundation**-certified Kubernetes distribution that is lightweight and focused, with options to install on Linux, Windows, and macOS.

Another example is K3s from Rancher, which is a Cloud Native Computing Foundation-certified Kubernetes distribution and is designed for production workloads running in resource-constrained environments such as IoT and edge computing deployments.

The minimal K3s Kubernetes cluster running on edge nodes is depicted in the following diagram:

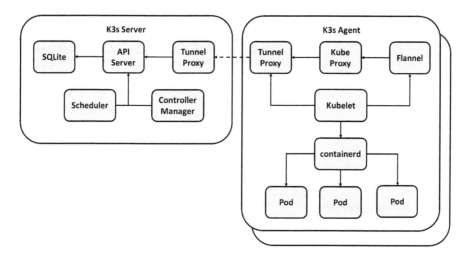

Figure 4.1 – K3s architecture

MicroK8s and K3s can be installed on public cloud virtual machines or even on a Raspberry Pi device. The architecture is highly optimized for unattended, remote installations on resource-constrained devices while preserving complete compatibility and compliance with Cloud Native Computing Foundation Kubernetes conformance tests.

By making it accessible and lightweight, MicroK8s and K3s are bringing Kubernetes to the edge computing layer.

A quick comparison of MicroK8s and K3s is shown in the following table:

Parameters	MicroK8s	Rancher K3s
Runtime	Native (snap)	Native
Supported architectures	amd64, arm64	x86_64, armhf, and arm64
Supported container runtimes	containerd	Docker, containerd
Startup time initial/following	~0:30 / ~0:30	
Memory requirements	1 GB/4 GB memory recommended	512 MB/1 GB memory recommended
Requires root?	Yes (rootless experimental)	
Multi-node support	Yes	Yes
Air-gap support	No	Yes

Table 4.1 – Comparison of MicroK8s and K3s

Optionally, you can also use platforms such as Google Anthos or AKS to manage and orchestrate container workloads on multiple clusters like the one shown here:

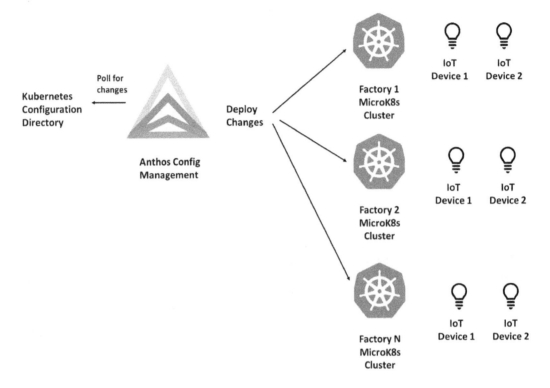

Figure 4.2 – Google Anthos on the cloud and MicroK8s at the edge

In the following chapters, we'll look at implementation aspects of common edge computing applications using MicroK8s.

Deployment of Kubernetes nodes at the edge

The core Kubernetes cluster is installed at a cloud provider or in your data center in this approach, with Kubernetes nodes deployed at the edge nodes. This is more appropriate for use cases where the infrastructure at the edge is constrained.

KubeEdge, an open source application that extends native containerized application orchestration and device management to hosts at the edge, is an example of this approach. KubeEdge is made up of two parts: the cloud and the edge.

It is based on Kubernetes and enables networking, application deployment, and metadata synchronization between the cloud and edge infrastructures. Developers can also use MQTT to write custom logic and enable resource-constrained device communication at the edge. KubeEdge is reliable and supports the most common IoT and edge use cases. It can be run on a compatible Linux distribution or an ARM device such as a Raspberry Pi.

The architecture of KubeEdge is shown here:

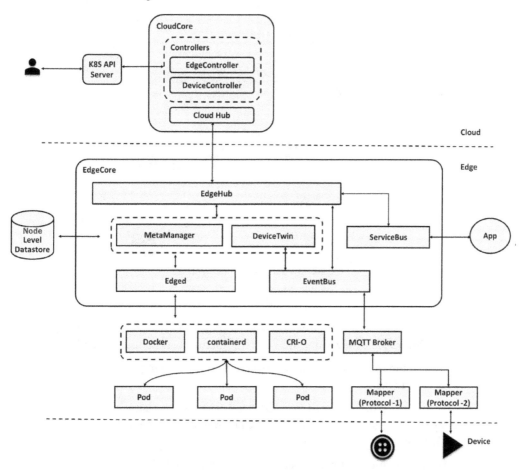

Figure 4.3 – KubeEdge architecture

Here's a quick rundown of KubeEdge's different components:

Component	Description
Edged	Containerized application management agent that runs on edge nodes.
EdgeHub	For edge computing, a websocket client is responsible for interfacing with the cloud service. This includes notifying edge-side host and device status changes to the cloud and syncing cloud-side resource modifications to the edge.
CloudHub	The web socket server is in charge of monitoring cloud updates, caching, and sending messages to EdgeHub.
EdgeController	An expanded Kubernetes controller that controls edge nodes and pod metadata, allowing data to be targeted to a specific edge node.
EventBus	The MQTT client for interacting with MQTT servers, allowing other components to publish and subscribe.
DeviceTwin	This is in charge of saving the device status and synchronizing it to the cloud. It also supports application query interfaces.
MetaManager	Between edged and EdgeHub, MetaManager is a message processor. It's also in charge of storing and retrieving metadata from and to a lightweight database (SQLite).

Table 4.2 – KubeEdge components

The following are some of KubeEdge's primary features:

- Users can orchestrate apps, manage devices, and monitor app and device status on edge nodes using KubeEdge, just like they can with a regular Kubernetes cluster in the cloud.

- Bidirectional communication, able to talk to edge nodes located in private subnets. Supports both metadata and data.

- Even when the edge is disconnected from the cloud, it can operate autonomously. Metadata is persistent per node; thus, no list-watching is required during node recovery. This allows you to get ready faster.

- At the edge, resource use is optimized. The memory footprint has been reduced to around 70 MB.

- For IoT and Industrial IoT, connectivity between applications and devices is simplified.

- Native support for x86, ARMv7, and ARMv8 architectures.

- Support for running third-party plugins and apps that rely on Kubernetes APIs on edge nodes with an autonomous Kube-API endpoint at the edge.

Deployment of virtual Kubernetes nodes at the edge

Virtual node agents reside in the cloud in this approach, but the abstract of nodes and Pods is deployed at the edge. Edge nodes carrying containers are given command control via virtual node agents.

Although there are others, Microsoft's Virtual Kubelet project is a nice example of a kubelet agent with a Kubernetes API extension. Virtual Kubelet is a Kubernetes agent that runs in a remote environment and registers itself as a cluster node. To build a node resource on the cluster, the agent uses the Kubernetes API. It uses the notions of taints and tolerations to schedule Pods in an external environment by calling its native API:

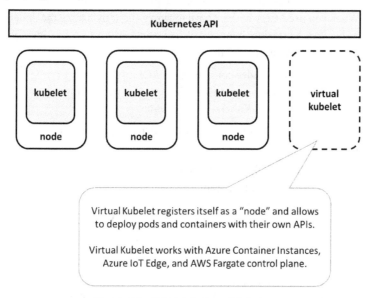

Figure 4.4 – Microsoft Virtual Kubelet

Virtual Kubelet works with providers such as AWS Fargate control plane, Azure Container Instances, and Azure IoT Edge. Following is a list of current providers (as of this writing).

Current list of Virtual Kubelet providers

In this section, we'll take a look at some of the Virtual Kubelet providers:

- **AWS Fargate**: Your Kubernetes cluster is connected to an AWS Fargate cluster using the AWS Fargate virtual-kubelet provider. The Fargate cluster appears as a virtual node with the CPU and memory resources you choose

  Pods scheduled on the virtual node execute on Fargate in the same way as they would on a regular Kubernetes node.

- **Admiralty Multi-Cluster Scheduler**: Admiralty is a Kubernetes controller system that schedules workloads intelligently among clusters. It's easy to use and integrates with other applications.

- **Alibaba Cloud Elastic Container Instance** (**ECI**): The Alibaba ECI provider is an adaptor that connects your Kubernetes cluster to the ECI service, allowing Pods from a K8s cluster to be implemented on Alibaba's cloud platform.

- **Azure Batch**: Azure Batch provides a distributed HPC computing environment on Azure. Azure Batch is a service that manages the scheduling of discrete processes and tasks across pools of virtual machines. It's frequently used for batch processing jobs such as rendering.

- **Azure Container Instances** (**ACI**): ACI in Azure provides a hosted environment for running containers. When you use ACI, you don't have to worry about managing the underlying compute infrastructure because Azure does it for you. When using ACI to run containers, you are charged per second for each container that is running

  The Virtual Kubelet's ACI provider configures an ACI instance as a node in any Kubernetes cluster. Pods can be scheduled on an ACI instance as if it were a conventional Kubernetes node when utilizing the Virtual Kubelet ACI provider

  This setup enables you to benefit from both Kubernetes' capabilities and ACI's management value and cost savings.

- **Elotl Kip**: Kip is a Virtual Kubelet provider that enables a Kubernetes cluster to launch Pods on their own cloud instances in a transparent manner. The kip Pod runs on a cluster and creates a virtual Kubernetes node within it

  Kip starts a right-sized cloud instance for the Pod's workload and sends the Pod to the instance when a Pod is scheduled on the Virtual Kubelet. The cloud instance is terminated once the Pod has finished operating. These cloud instances are referred to as *cells*

  When workloads run on Kip, your cluster size naturally scales with the cluster workload, Pods are strongly isolated from each other, and the user is freed from managing worker nodes and strategically packing Pods onto nodes. This results in lower cloud costs, improved security, and simpler operational overhead.

- **Kubernetes Container Runtime Interface** (**CRI**): The CRI provider implementation should be regarded as a bare-bones minimal implementation for testing the Virtual Kubelet project's core against real Pods and containers; in other words, it is more extensive than MockProvider

  This provider implementation is also built in such a way that it may be used to prototype new architectural features on local Linux infrastructure. If the CRI provider can be demonstrated to run effectively within a Linux guest, it may be assumed that the abstraction will work for other providers as well.

- **Huawei Cloud Container Instance** (**CCI**): The Huawei CCI virtual kubelet provider sets up a CCI project as a node in any Kubernetes cluster, including **Huawei Cloud Container Engine** (**CCE**). As a private cluster, CCE provides native Kubernetes applications and tools, allowing you to quickly build up a container runtime environment. The virtual kubelet provider's scheduled Pod will run in CCI, taking advantage of CCI's high performance.

- **HashiCorp Nomad**: By exposing the Nomad cluster as a node in Kubernetes, the HashiCorp Nomad provider for Virtual Kubelet connects your Kubernetes cluster with the Nomad cluster. Pods scheduled on the virtual Nomad node registered on Kubernetes will run as jobs on Nomad clients, just like they would on a Kubernetes node, if you use the provider.

- **Liqo**: Liqo is an on-prem or managed platform that enables dynamic and decentralized resource sharing among Kubernetes clusters. Liqo makes it possible to launch Pods on a distant cluster without modifying Kubernetes or the apps

  With Liqo, you can extend a Kubernetes cluster's control plane across cluster boundaries, making multi-clusters native and transparent: collapse a complete remote cluster to a virtual local node, allowing task offloading and resource management in accordance with standard Kubernetes practices.

- **OpenStack Zun**: Your Kubernetes cluster is connected to an OpenStack cloud through the OpenStack Zun virtual kubelet provider. Because each Pod is provided with dedicated Neutron ports in your tenant subnets, your OpenStack Pods have access to OpenStack tenant networks.

- **Tencent Games tensile-kube**: This allows Kubernetes clusters to collaborate. Tensile-kube is based on Virtual Kubelet and offers the following features:

  - Cluster resources are discovered automatically.

  - Pods are notified of changes in real time, reducing the expense of frequent lists.

  - All kubectl logs and kubectl exec operations are supported.

  - When utilizing a multi-scheduler, schedule Pods globally to avoid unscheduled Pods owing to resource fragmentation.

  - If Pods can't be scheduled in lower clusters, use a descheduler to reschedule them.

  - Supports PV/PVC and service abstractions.

Deployment of Kubernetes devices at the edge

The Kubernetes device plugin framework is used to expose leaf devices as resources in a Kubernetes cluster in this approach.

This technique is demonstrated by **Microsoft Akri**, which exposes a variety of sensors, controllers, and MCU class leaf devices as resources in a Kubernetes cluster. The Akri project brings the Kubernetes device plugin concept to the edge, where a variety of leaf devices use different communication protocols and have sporadic availability:

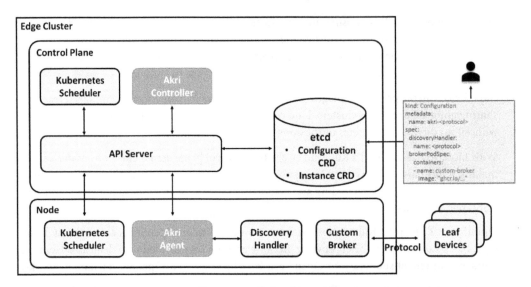

Figure 4.5 – Akri architecture

ONVIF, Udev, and OPC UA discovery handlers are now supported by Akri. More protocol support is being developed.

As seen previously, a variety of deployment approaches indicate how Kubernetes may be utilized for edge workloads, as well as support for the architecture that fulfills enterprise application needs such as low latency, resource constraints, data privacy, and bandwidth scalability, among others.

In the next section, we will look at how Kubernetes is suitable for running edge workloads.

Propositions that Kubernetes offers

In terms of resource and workload control, edge-based infrastructure poses various challenges. There would be thousands of edge nodes and distant edge nodes to control in a short amount of time. Organizations' edge architecture is designed to provide more centralized autonomy from the cloud, security standards, and relatively low latency. Take a peek at what Kubernetes for edge has to offer:

Requirement	Propositions that Kubernetes offer
Resource constraints: Low compute capability, small device footprints.	For production workloads running in extremely limited contexts such as IoT and edge computing deployments, certified lightweight versions of Kubernetes distributions (for example, MicroK8s and K3s) are available.
Security challenges: Data privacy, physical device security, and network security of the connected devices.	Provides a policy-based framework for all deployment types, allowing any policies or rule set(s) to be applied to the total infrastructure. Policies can also be tailored to specific channels or edge nodes based on configuration requirements.
Manageability: Manage application software across thousands of devices from many different suppliers.	The use of **Custom Resource Definition** (**CRD**) and Operators allows site reliability engineering teams to manage complex applications and infrastructure with ease and power.
Reliability: Consistency in building, deployment, and maintenance of applications.	The GitOps and DevOps pipelines-based strategy provides system complexity management by pushing only vetted changes to configuration settings and application artifacts into operational systems.
Automation: Provision for automated means to deploy and manage multiple distributed applications across any number of machines, physical or virtual, with high levels of automation.	The Kubernetes cluster is continually reconciling to a point of target state that developers or administrators have established. Kubernetes employs a pipeline method, in which YAML-based configuration and container image modifications are first committed to Git, followed by pipeline operations, which result in updates to apps and the cluster as a whole.

Table 4.3 – Propositions that Kubernetes offers

Kubernetes is a critical component of businesses that are evolving into digital-first enterprises. Kubernetes is currently being deployed in 59% of data centers to increase resource efficiency and offer agility to the software development cycle, according to reports. In a distributed cloud environment, Kubernetes can manage and orchestrate containerized applications as well as legacy virtual machines. Kubernetes can also be used to execute AI/ML and GPU workloads, in addition to pure applications.

Kubernetes is certainly the go-to platform for edge computing, at least for those edge contexts that require dynamic orchestration for apps and centralized administration of workloads, according to the Linux Foundation's *The State of the Edge Report*. Kubernetes extends the benefits of cloud-native computing software development to the edge, allowing for flexible and automated management of applications that span a disaggregated cloud environment.

By deploying and testing Kubernetes at the edge, enterprises and telecom operators can achieve a high level of flexibility, observability, and dynamic orchestration.

Summary

As companies embrace digital transformation, Industry 4.0, industrial automation, smart manufacturing, and all the advanced use cases that these initiatives provide, the relevance of Kubernetes, edge, and cloud collaborating to drive intelligent business decisions is becoming clear. We've looked at a different approach that shows how Kubernetes may be used for running edge workloads. In the next chapters, we'll go over the deployment of a whole Kubernetes cluster at the edge approach in depth. Other approaches are beyond the scope of this book.

In the following chapters, we'll go over implementation aspects of common edge computing applications using MicroK8s in detail, such as running your applications on a multi-node Raspberry Pi cluster; configuring load balancing; installing/configuring different CNI plugins for network connectivity; configuring logging, monitoring, and alerting options; and building/deploying ML models and serverless applications.

Also, we will look into setting up storage replication for your stateful applications, implementing a service mesh for cross-cutting concerns, high-availability clusters to withstand a component failure and continue to serve workloads without interruption, the configuration of containers with workload isolation, and running secured containers with isolation from the host system.

Part 3:
Running Applications
on MicroK8s

This part focuses on the implementation aspects that are common for any IoT/edge computing applications, such as running your applications on a multi-node Raspberry Pi cluster, installing/configuring different CNI plugins for network connectivity, configuring load balancing, configuring logging, monitoring, alerting options, building, and deploying machine learning models, and serverless applications.

This part of the book comprises the following chapters:

- *Chapter 5, Creating and Implementing Updates on Multi-Node Raspberry Pi Kubernetes Clusters*

- *Chapter 6, Configuring Connectivity for Containers*

- *Chapter 7, Setting Up MetalLB and Ingress for Load Balancing*

- *Chapter 8, Monitoring the Health of Infrastructure and Applications*

- *Chapter 9, Using Kubeflow to Run AI/MLOps Workloads*

- *Chapter 10, Going Serverless with Knative and OpenFaaS Frameworks*

## 5

Creating and Implementing Updates on a Multi-Node Raspberry Pi Kubernetes Clusters

Companies are embracing digital transformation, Industry 4.0, industrial automation, smart manufacturing, and all the advanced use cases that these initiatives provide, as we saw in the previous chapter. As a result, the importance of Kubernetes, edge, and cloud collaboration to drive intelligent business decisions is becoming clear. Kubernetes is steadily becoming the go-to platform for edge computing and extends the benefits of cloud-native technologies to the edge, allowing for the flexible and automated management of applications that span a disaggregated cloud environment. In this and the following chapters, we will be looking at implementation aspects of common edge computing applications using the MicroK8s Kubernetes platform.

Reiterating the points that we discussed in the previous chapter, **Canonical MicroK8s** (`https://microk8s.io/`) is a powerful, **Cloud Native Computing Foundation** (**CNCF**)-certified Kubernetes distribution. Here are some of the key reasons why it has become a powerful enterprise computing platform:

- **Delivered as a snap package**: These are application packages for desktop, cloud, and even **Internet of Things** (**IoT**) devices that are simple to install, secured with auto-updates, and can be deployed on any Linux distributions that support snaps.
- **Strict confinement**: This ensures complete isolation from the underlying **operating system** (**OS**) as well as a highly secure Kubernetes environment fit for production.

- **Production-grade add-ons**: Add-ons such as Istio, Knative, CoreDNS, Prometheus, Jaeger, Linkerd, Cilium, and Helm are available. They are straightforward to set up, requiring only a few lines of commands. For better **artificial intelligence (AI)** and **machine learning (ML)** capabilities, Kubeflow is also available as an add-on to MicroK8s.
- **Automatic, autonomous, and self-healing high availability (HA)**: For clusters of three or more Nodes, MicroK8s automatically activates HA. With no administrative intervention, MicroK8s provides resilient and self-healing HA.

In this chapter, we're going to cover the following main topics:

- Creating a MicroK8s multi-node cluster using a Raspberry Pi
- Deploying a sample containerized application
- Performing rolling updates to the application with a new software version
- Scaling the application deployment
- Guidelines on multi-node cluster configuration
- Container life cycle management
- Deploying and sharing HA applications

Creating a MicroK8s multi-node cluster using a Raspberry Pi

Before we delve into the steps on how to create a MicroK8s multi-node cluster, let's recap the key concepts of Kubernetes that we covered earlier, as follows:

1. A **Kubernetes cluster** (like the one shown in *Figure 5.1*) would have the following two types of resources:

    a. A **control plane** that controls the cluster

    b. **Nodes**—the worker Nodes that run applications

2. All actions in the cluster are coordinated by the control plane, including scheduling applications, maintaining the intended state of applications, scaling applications, and rolling out new updates, among other things. Each node can be a **virtual machine (VM)** or a physical computer that serves as a worker machine in a cluster.

3. A **Kubernetes control plane** is a collection of three processes: an **application programming interface (API) server**, a **controller manager**, and a **scheduler**.

4. Each individual non-control plane node on the cluster has a **kubelet** process that takes care of the communication with the Kubernetes control plane, the **kube-proxy** process for all network communications, and a **container runtime** such as Docker.

5. The control plane issues a command to start the application containers when any applications need to be deployed on Kubernetes. Containers are scheduled to run on the cluster's Nodes by the control plane.

6. The Nodes communicate with the control plane using the Kubernetes API, which the control plane exposes.

A typical Kubernetes architecture, system, and abstractions are shown in the following diagram:

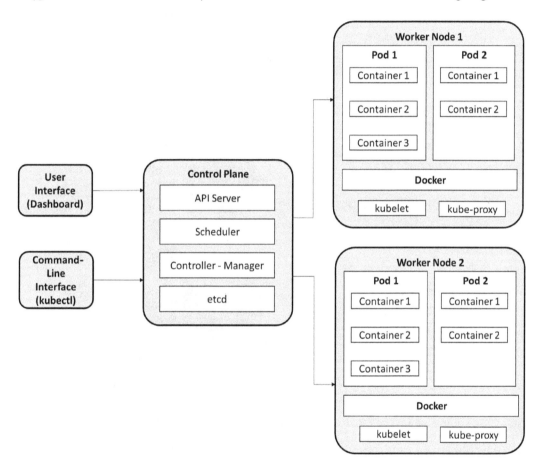

Figure 5.1 – Kubernetes system and abstractions

Now that we are clear on the Kubernetes architecture, system, and abstractions, we will delve into the steps of creating a Kubernetes Raspberry Pi multi-node cluster.

What we are trying to achieve

We'll list down the steps that we're seeking to accomplish in this section. Prepare Raspberry Pi 4 boards for MicroK8s installation by doing the following:

1. Installing and configuring MicroK8s on each of the boards

2. Adding Nodes to the cluster

3. Joining multiple deployments to form a two-node cluster (one control plane node/one worker node)

The Raspberry Pi cluster that we will build in this step is depicted in the following diagram:

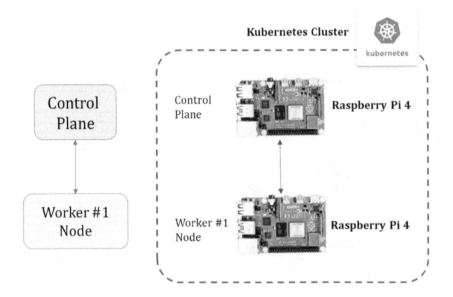

Figure 5.2 – What we are trying to achieve

Now that we know what we want to do, let's look at the requirements.

Before you begin, here are the prerequisites for building a Raspberry Pi Kubernetes cluster:

- A microSD card (4 **gigabytes** (**GB**) minimum; 8 GB recommended)

- A computer with a microSD card drive

- A Raspberry Pi 2, 3, or 4 (one or more)

- A **micro-Universal Serial Bus** (**micro-USB**) power cable (USB-C for the Pi 4)

- A Wi-Fi network or an Ethernet cable with an internet connection

- (Optional) A monitor with a **High-Definition Multimedia Interface** (**HDMI**) interface
- (Optional) An HDMI cable for the Pi 2 and 3 and a micro-HDMI cable for the Pi 4
- (Optional) A USB keyboard

Now that we've established the requirements, we'll go on to the step-by-step instructions on how to complete the process.

Step 1a: Installing OS image onto SD card

The first step is to install an OS image onto the microSD card. To do that, we will be using the **Raspberry Pi Imager tool**, as shown in the following screenshot, to install an OS image to a microSD card that can then be used with Raspberry Pi:

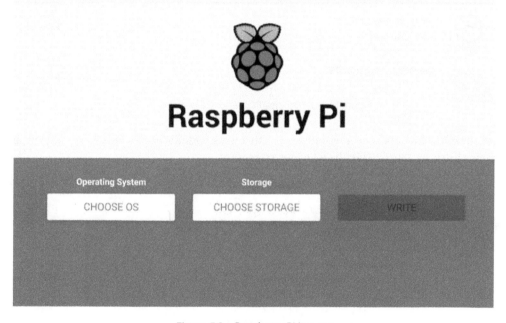

Figure 5.3 – Raspberry Pi Imager

Download and install **Raspberry Pi Imager** from the Raspberry Pi website on a computer equipped with a **Secure Digital** (**SD**) card reader. Run Raspberry Pi Imager with the microSD card you'll be using and open the **CHOOSE OS** menu.

Choose **Other general purpose OS** from the options listed, as illustrated in the following screenshot:

Figure 5.4 – Raspberry Pi Imager OS options

Choose any of the **Ubuntu Server** 64-bit versions that work with Raspberry Pi 2, 3, 3+, and 4 from the options listed, as illustrated in the following screenshot:

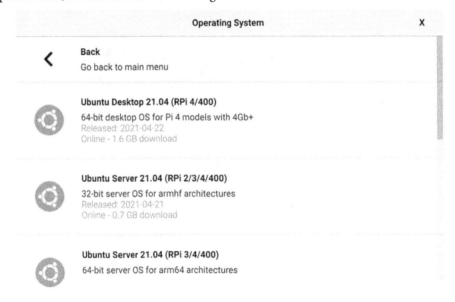

Figure 5.5 – Choosing any Ubuntu Server version that works with Raspberry Pi Imager 2/3/4

> **Note**
> MicroK8s is only available for 64-bit Ubuntu images.

Open the **SD Card** menu after selecting an image. Choose the microSD card that you've inserted. Click **WRITE** to start the operation and Raspberry Pi Imager will wipe your microSD card data. You will be prompted to confirm this procedure.

The process is illustrated in the following screenshot:

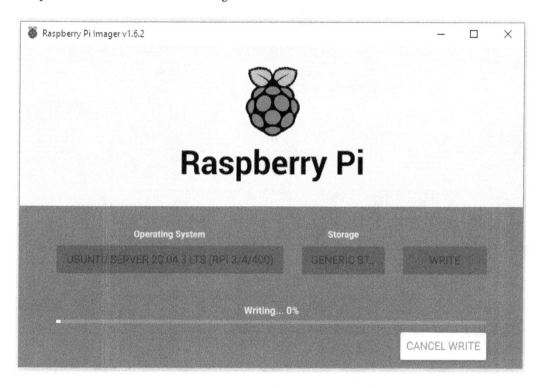

Figure 5.6 – Raspberry Pi Imager write operation

Post confirmation, Raspberry Pi Imager will start flashing OS images to the microSD card. It will take a while to finish. The message that displays once finished is illustrated in the following screenshot:

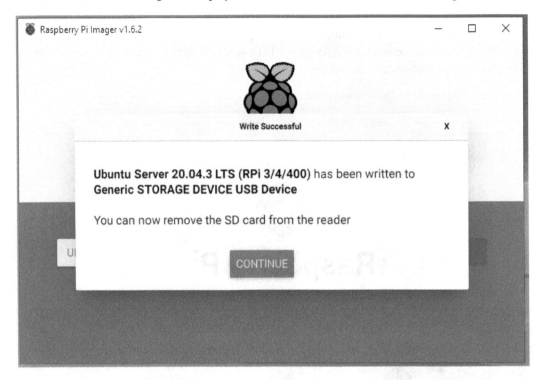

Figure 5.7 – Raspberry Pi Imager: write operation completed

Once finished, continue with the configuration of Wi-Fi access, remote access, control groups, and hostname settings.

Configuring Wi-Fi access settings

Open a file manager—as shown in the following screenshot—while the SD card is still inserted in your laptop, and look for the system-boot partition on the card. It holds the initial configuration files that are loaded during the first boot:

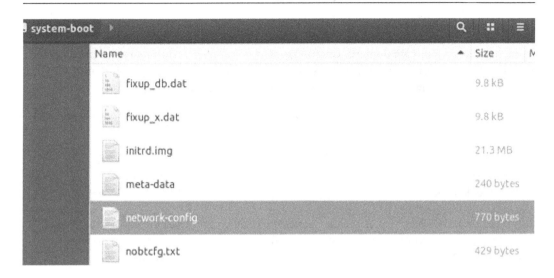

Figure 5.8 – Configuring Wi-Fi access settings

Modify the network-config file to include your Wi-Fi credentials. Here is an example of Wi-Fi configuration:

```
wifis:
  wlan0:
    dhcp4: true
    optional: true
    access-points:
      "Karthik Home":
        password: "Karthik123"
```

We've finished configuring Wi-Fi access and are ready to go on to the next step.

Step 1b: Configuring remote access settings

By default, the **Secure Shell** (**SSH**) network protocol is disabled. To be able to connect to your Raspberry Pi remotely, you need to enable it explicitly. Open the system-boot partition on the card and create an empty file named ssh without a file extension. During the boot process, it will automatically prepare and set up SSH on your Raspberry Pi if it detects this file.

We've finished configuring remote access and are ready to go on to the next step of configuring control groups.

Step 1c: Configuring control group settings

By default, control groups are disabled. Control groups (abbreviated as **cgroups**) are a Linux kernel feature that limits, accounts for, and isolates the resource usage (**central processing unit (CPU)**, memory, disk **input/output (I/O)**, network, and so on) of a collection of processes.

Open the configuration file located at `/boot/firmware/cmdline.txt` on the card and add the following options:

```
cgroup_enable=memory cgroup_memory=1
```

A full line would look like this:

```
cgroup_enable=memory cgroup_memory=1 net.ifnames=0 dwc_otg.
lpm_enable=0 console=ttyAMA0,115200 console=tty1 root=/dev/
mmcblk0p2 rootfstype=ext4 elevator=deadline rootwait
```

Save the file, extract the card from your laptop, and insert it into the Raspberry Pi. Before powering the Pi, connect an HDMI screen and a USB keyboard. Power on the Pi, and you will be able to see the boot process on the screen. It typically takes less than 2 minutes to complete the booting process.

Once the boot is finished, connect to your Raspberry Pi using any SSH client (for example, `putty`), and continue the following configuration:

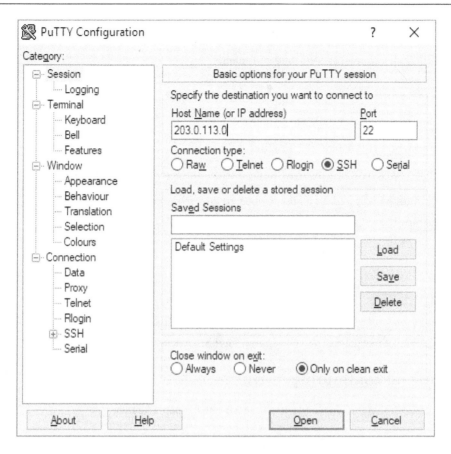

Figure 5.9 – PuTTY SSH client

Type in the IP address of the Raspberry Pi and click **Open** to connect. On the first connect, you will be asked to confirm the connection: click **Accept** to confirm.

On the login page, type ubuntu as both the username and the password. Ubuntu will ask you to change your password to something else. After that, use the ssh command and the new password to reconnect.

Success! You are now connected to Ubuntu Server running on your Raspberry Pi.

We've finished configuring most of the settings and are ready to go on to the next step of hostname configuration.

Step 1d: Configuring hostname

For Ubuntu OS, by default, the hostname would be ubuntu, but since we need different hostnames to be identified in the cluster, we would need to change this based on our needs. For the cluster we are creating, I'm going to name one of the Nodes controlplane and the others WorkerXX.

Follow the next steps to change the hostname.

From the `putty` shell, enter the following command:

```
sudo nano /etc/hostname
```

Modify the hostname and exit the `nano` editor using *Ctrl + X*.

Type `sudo reboot` for changes to take effect.

Congratulations! You have now completed all the configurations for your Raspberry Pi.

Steps 1a to *1d* must be repeated for all the Raspberry Pis in the cluster.

Post successful boot, ensure your packages are updated to the latest version, and run the following commands:

```
sudo apt update
sudo apt upgrade
```

We've finished configuring all the settings for all the Raspberry Pis on the cluster and are ready to go on to the next step of installation and configuration of MicroK8s.

Installing and configuring MicroK8s

SSH into your control plane node and install the MicroK8s snap, like this:

```
sudo snap install microk8s --classic
```

The following command execution output confirms that the MicroK8s snap has been installed successfully:

```
ubuntu@controlplane:~$ sudo snap install microk8s --classic
microk8s (1.23/stable) v1.23.3 from Canonical√ installed
ubuntu@controlplane:~$ 
```

Figure 5.10 – MicroK8s snap installation successful

Now that we have installed the MicroK8s snap, type the `microk8s status` command to verify its running state, as follows:

```
ubuntu@controlplane:~$ microk8s status
Insufficient permissions to access MicroK8s.
You can either try again with sudo or add the user ubuntu to the 'microk8s' grou
p:

    sudo usermod -a -G microk8s ubuntu
    sudo chown -f -R ubuntu ~/.kube

After this, reload the user groups either via a reboot or by running 'newgrp mic
rok8s'.
```

Figure 5.11 – Verifying whether MicroK8s is running

As indicated in the preceding command execution output, join the user in the MicroK8s group and gain access to a `.kube` caching directory using the following set of commands:

```
sudo usermod -a -G microk8s ubuntu
sudo chown -f -R ubuntu ~/.kube
```

Retype the `microk8s status` command to verify whether it's running. The following command execution output confirms that MicroK8s is running successfully:

```
ubuntu@controlplane:~$ microk8s status
microk8s is running
high-availability: no
  datastore master nodes: 127.0.0.1:19001
  datastore standby nodes: none
addons:
  enabled:
    ha-cluster            # Configure high av
  disabled:
    dashboard             # The Kubernetes da
    dashboard-ingress     # Ingress definitio
```

Figure 5.12 – MicroK8s is running

Now that we have installed MicroK8s, the next step is to create a `kubectl` alias with the following command:

```
sudo snap alias microk8s.kubectl kubectl
```

The following command execution output confirms that an alias has been added successfully:

```
ubuntu@controlplane:~$ sudo snap alias microk8s.kubectl kubectl
Added:
  - microk8s.kubectl as kubectl
ubuntu@controlplane:~$ 
```

Figure 5.13 – kubectl alias successfully added

If the installation has been successful, you should then see the following output:

```
ubuntu@controlplane:~$ kubectl get nodes
NAME            STATUS    ROLES     AGE     VERSION
controlplane    Ready     <none>    12m     v1.23.3-2+0d2db09fa6fbbb
ubuntu@controlplane:~$ █
```

Figure 5.14 – Verifying whether the node is in a Ready state

Repeat the MicroK8s installation process on the other Nodes as well.

Here is the command execution output of the MicroK8s installation on the worker node:

```
ubuntu@worker1:~$ sudo snap install microk8s --classic
microk8s (1.23/stable) v1.23.3 from Canonical✓ installed
```

Figure 5.15 – MicroK8s snap installation on worker1 node successful

The following command execution output confirms that MicroK8s is running successfully on the worker node as well:

```
ubuntu@worker1:~$ microk8s status
microk8s is running
high-availability: no
  datastore master nodes: 127.0.0.1:19001
  datastore standby nodes: none
addons:
  enabled:
    ha-cluster              # Configure high av
  disabled:
    dashboard               # The Kubernetes da
    dashboard-ingress       # Ingress definitio
```

Figure 5.16 – Verifying whether MicroK8s is running

Now that MicroK8s is running, the next step is to check whether the kubectl get Nodes command displays the node in a Ready state, as indicated in the following command execution output:

```
ubuntu@worker1:~$ kubectl get nodes
NAME        STATUS    ROLES     AGE     VERSION
worker1     Ready     <none>    13m     v1.23.3-2+0d2db09fa6fbbb
ubuntu@worker1:~$ █
```

Figure 5.17 – Verifying whether the node is in a Ready state

We have completed the installation of MicroK8s on all boards. The next step is to add a worker node to the control plane node. Open the `putty` shell to the control plane node and run the following command to generate a connection string:

```
sudo microk8s.add-node
```

The following command execution output validates that the command was successfully executed and provides instructions for the connection string:

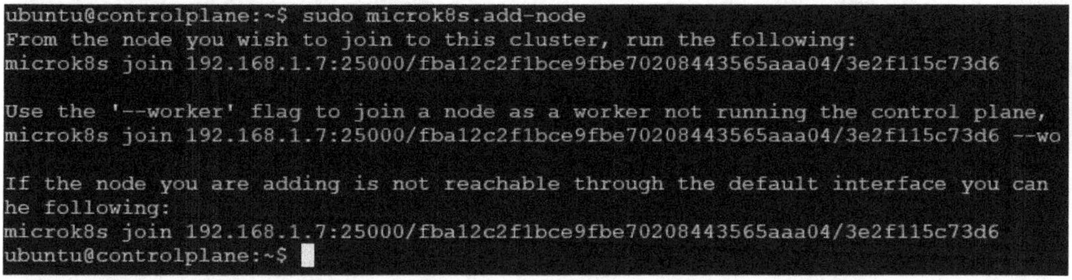

```
ubuntu@controlplane:~$ sudo microk8s.add-node
From the node you wish to join to this cluster, run the following:
microk8s join 192.168.1.7:25000/fba12c2f1bce9fbe70208443565aaa04/3e2f115c73d6

Use the '--worker' flag to join a node as a worker not running the control plane,
microk8s join 192.168.1.7:25000/fba12c2f1bce9fbe70208443565aaa04/3e2f115c73d6 --wo

If the node you are adding is not reachable through the default interface you can
he following:
microk8s join 192.168.1.7:25000/fba12c2f1bce9fbe70208443565aaa04/3e2f115c73d6
ubuntu@controlplane:~$
```

Figure 5.18 – Generating connection string for adding Nodes

As indicated by the preceding command execution output, a connection string is generated in the form of `<control plane_ip>:<port>/<token>`.

Adding the worker node

We now have the connection string to join with the control plane node. Open the `putty` shell to the worker node and run the `join` command to add it to the cluster, as follows:

```
microk8s join <control plane_ip>:<port>/<token>
```

The command was successfully executed, and the node has joined the cluster, as shown in the following output:

```
ubuntu@worker1:~$ sudo microk8s join 192.168.1.7:25000/fba12c2f1bce9fbe702084435
65aaa04/3e2f115c73d6 --worker
Contacting cluster at 192.168.1.7

The node has joined the cluster and will appear in the nodes list in a few secon
ds.

Currently this worker node is configured with the following kubernetes API serve
r endpoints:
    - 192.168.1.7 and port 16443, this is the cluster node contacted during the
join operation.

If the above endpoints are incorrect, incomplete or if the API servers are behin
d a loadbalancer please update
/var/snap/microk8s/current/args/traefik/provider.yaml

ubuntu@worker1:~$ 
```

Figure 5.19 – Adding worker1 node to the cluster

As indicated by the preceding command execution output, you should be able to see the new node in a few seconds on the control plane.

Use the following command to verify whether the new node has been added to the cluster:

```
kubectl get nodes
```

The following command execution output shows that controlplane and worker1 are part of the cluster:

```
ubuntu@controlplane:~$ kubectl get nodes
NAME            STATUS     ROLES      AGE     VERSION
controlplane    Ready      <none>     17m     v1.23.3-2+0d2db09fa6fbbb
worker1         Ready      <none>     27s     v1.23.3-2+0d2db09fa6fbbb
ubuntu@controlplane:~$ 
```

Figure 5.20 – Cluster is ready; controlplane and worker1 are part of the cluster

The completed cluster should resemble the one as shown here:

Figure 5.21 – Our cluster is ready

> **Note**
>
> If you want to remove a node from the cluster, run the following command on the control plane: `sudo microk8s remove-node <node name>`. Alternatively, you can leave the cluster from the worker node by running `sudo microk8s.leave`.

At this point, you have a fully functional multi-node Kubernetes cluster. To summarize, we have installed MicroK8s on the Raspberry Pi boards and joined multiple deployments to form the cluster. We've seen how to add Nodes to the cluster as well. In the next section, we are going to deploy a sample application on the MicroK8s cluster we just created.

Deploying a sample containerized application

In this section, we will be deploying the following nginx deployment from the Kubernetes examples repository on our multi-node MicroK8s cluster setup:

```
apiVersion: apps/v1
kind: Deployment
metadata:
  name: nginx-deployment
spec:
  selector:
    matchLabels:
      app: nginx
```

```
  replicas: 2 # tells deployment to run 2 pods matching the
template
  template:
    metadata:
      labels:
        app: nginx
    spec:
      containers:
      - name: nginx
        image: nginx:1.14.2
        ports:
        - containerPort: 80
      affinity:
        podAntiAffinity:
          requiredDuringSchedulingIgnoredDuringExecution:
          - labelSelector:
              matchExpressions:
              - key: "app"
                operator: In
                values:
                - nginx
            topologyKey: "kubernetes.io/hostname"
```

The following command will deploy the previous sample application deployment:

```
kubectl apply -f deployment.yaml
```

The following command execution output indicates that there is no error in the deployment, and in the next step, we can verify this using the describe command:

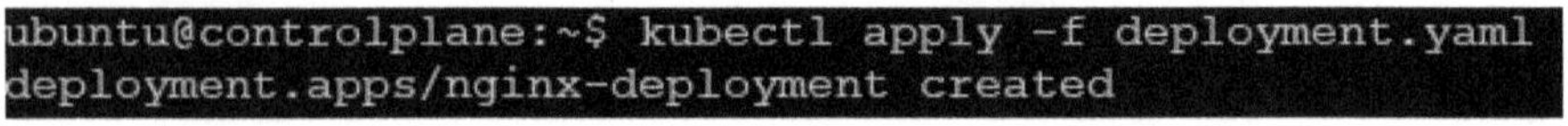

```
ubuntu@controlplane:~$ kubectl apply -f deployment.yaml
deployment.apps/nginx-deployment created
```

Figure 5.22 – Sample application deployment

The following command execution output displays information about the deployment:

```
ubuntu@controlplane:~$ kubectl describe deployment nginx-deployment
Name:                   nginx-deployment
Namespace:              default
CreationTimestamp:      Sat, 05 Mar 2022 12:11:03 +0000
Labels:                 <none>
Annotations:            deployment.kubernetes.io/revision: 1
Selector:               app=nginx
Replicas:               2 desired | 2 updated | 2 total | 2 available | 0 unavaila
StrategyType:           RollingUpdate
MinReadySeconds:        0
RollingUpdateStrategy:  25% max unavailable, 25% max surge
Pod Template:
  Labels:  app=nginx
  Containers:
   nginx:
    Image:         nginx:1.14.2
    Port:          80/TCP
    Host Port:     0/TCP
    Environment:   <none>
    Mounts:        <none>
  Volumes:         <none>
Conditions:
  Type           Status  Reason
  ----           ------  ------
  Available      True    MinimumReplicasAvailable
  Progressing    True    NewReplicaSetAvailable
OldReplicaSets:  <none>
NewReplicaSet:   nginx-deployment-57d554699f (2/2 replicas created)
Events:
  Type    Reason            Age    From                   Message
  ----    ------            ----   ----                   -------
  Normal  ScalingReplicaSet 2m5s   deployment-controller  Scaled up replica set ng
ubuntu@controlplane:~$ █
```

Figure 5.23 – Describing sample application deployment

Check the pods' status to verify whether the application has been deployed and running, as follows:

```
kubectl get pods -l app=nginx
```

The following command execution output indicates that pods have been created and that their status is Running:

```
ubuntu@controlplane:~$ kubectl get pods -l app=nginx
NAME                              READY   STATUS    RESTARTS   AGE
nginx-deployment-57d554699f-clxd5  1/1    Running   0          3m9s
nginx-deployment-57d554699f-8hjbv  1/1    Running   0          3m9s
ubuntu@controlplane:~$ █
```

Figure 5.24 – Checking whether pods are in a Running status

Let's also check where the pods are running using the following command:

```
kubectl get pods -l app=nginx -o wide
```

The following command execution output indicates that pods are equally distributed between the Nodes:

```
ubuntu@controlplane:~$ kubectl get pods -l app=nginx -o wide
NAME                                 READY   STATUS    RESTARTS  AGE     IP             NODE
S
nginx-deployment-57d554699f-clxd5    1/1     Running   0         8m56s   10.1.49.70     controlplane
nginx-deployment-57d554699f-8hjbv    1/1     Running   0         8m56s   10.1.235.129   worker1
ubuntu@controlplane:~$
```

Figure 5.25 – Checking whether pods are equally distributed

Great! We have just deployed our sample application deployment on the Raspberry Pi multi-node cluster. Here is what Kubernetes has done for us:

- Looked for a suitable node on which to run an instance of the application (we have two available Nodes) and scheduled the application to run on that node based on podAntiAffinity rules.

- podAntiAffinity rules limit the pod deployments on which Nodes the pod can be scheduled based on labels from other pods currently operating on the node.

- Configured the cluster to reschedule the instance on a new node when needed.

Pod topology spread constraints can also be used to regulate how pods are distributed among failure domains such as regions, zones, Nodes, and other user-defined topology domains in your cluster. This can aid in achieving HA and resource-use efficiency.

To summarize, we built a Kubernetes Raspberry Pi cluster and used it to deploy a sample application. In the next step, we'll perform rolling updates to the application we've just deployed.

Performing rolling updates to the application with a new software version

The rolling updates feature of Kubernetes allows Deployments to be updated with zero downtime. It handles the upgrading of pods' instances with new ones in an incremental manner, and new pods would be scheduled on Nodes that have resources available.

Some key features of rolling updates are listed here:

- Transferring an application from one environment to another (via container image updates).

- Rollback to a prior version of the application.

- With minimal downtime, **continuous integration** and **continuous delivery** (**CI/CD**) of applications are achievable.

We are going to reuse the same example of the nginx sample deployment that we used earlier. We can update the same deployment by applying the following new YAML file.

This YAML file specifies that the deployment should be updated to use the nginx 1.16.1 container image instead of nginx.1.14.2:

```
apiVersion: apps/v1
kind: Deployment
metadata:
  name: nginx-deployment
spec:
  selector:
    matchLabels:
      app: nginx
  replicas: 2
  strategy:
    rollingUpdate:
      maxSurge: 0
      maxUnavailable: 1
    type: RollingUpdate
  template:
    metadata:
      labels:
        app: nginx
    spec:
      containers:
      - name: nginx
        image: nginx:1.16.1 # Update the version of nginx from
1.14.2 to 1.16.1
        ports:
        - containerPort: 80
      affinity:
        podAntiAffinity:
          requiredDuringSchedulingIgnoredDuringExecution:
          - labelSelector:
              matchExpressions:
              - key: "app"
                operator: In
                values:
```

```
      - nginx
  topologyKey: "kubernetes.io/hostname"
```

The following command will deploy the preceding updated application deployment that uses the nginx 1.16.1 image instead of the nginx 1.14.2 image:

```
kubectl apply -f deployment-update.yaml
```

The following output after the execution of the previous command confirms that there is no error in the deployment, and in the next steps, we can verify the recreation of pods with new names and delete the old pods:

```
ubuntu@controlplane:~$ kubectl apply -f deployment-update.yaml
deployment.apps/nginx-deployment configured
ubuntu@controlplane:~$
```

Figure 5.26 – Update to sample application deployment

The following command execution output indicates that pods have been recreated and that their status is Running:

```
ubuntu@controlplane:~$ kubectl get pods -l app=nginx
NAME                                READY   STATUS    RESTARTS   AGE
nginx-deployment-6796bd85dd-1f4zb   1/1     Running   0          7m36s
nginx-deployment-6796bd85dd-6fgtp   1/1     Running   0          95s
```

Figure 5.27 – New pods of updated deployment

Here is how a rolling update works (refer to *Figure 5.29*):

1. Using the revised configuration, it creates a new deployment.

2. Increases/decreases the number of replicas on the new and old controllers until the correct number is attained.

3. Finally, the original deployment and associated pods will be deleted.

Here is a diagram of rolling updates' functionality:

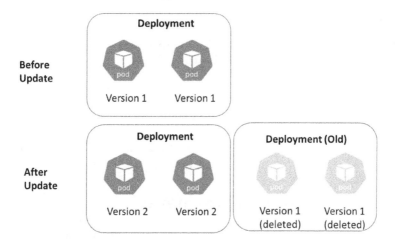

Figure 5.28 – Rolling updates to the sample application

There are two additional options when utilizing the RollingUpdate approach to fine-tune the update process, as follows:

- maxSurge: During an update, the maximum number of pods that can be created is greater than the desired number of pods.

- maxUnavailable: The number of pods that may become unavailable during the upgrade procedure.

We've set maxSurge to 0 and maxUnavailable to 1 in the sample application deployment, indicating that the maximum number of new pods that can be generated at a time is 0 and the maximum number of old pods that can be destroyed at a time is 1. This strategy indicates that as new pods are created, old pods will be destroyed one by one.

Depending on your goal, there are various sorts of deployment tactics you might want to use. For example, you may need to deploy modifications to a specific environment for more testing, or to a group of users/customers, or you may wish to do user testing.

Various Kubernetes deployment strategies are outlined here:

- **Recreate**: In this very simple deployment strategy, all old pods are killed at the same time and replaced with new ones.

- **Blue/green deployments**: In a blue/green deployment strategy, the old (green) and new (blue) versions of the application are deployed at the same time. When both are launched, consumers may only access the green deployment; the blue deployment is available to your **quality assurance (QA)** team for test automation on a different service or via direct port forwarding.

- **Canary deployments**: Canary deployments are like blue/green deployments, except they are more controlled and use a *progressive delivery* phased-in technique. Canary encompasses a variety of methods, including dark launches and A/B testing.

- **Dark deployments or A/B deployments**: Another variation on a canary deployment is a dark deployment. The distinction between dark and canary deployments is that dark deployments deal with features on the frontend rather than the backend, as canaries do.

To summarize, we've launched a sample application as well as performed rolling updates on the one we've already deployed. We will concentrate on how to scale the application in the following section.

Scaling the application deployment

Changing the number of replicas in a Deployment allows scaling the deployments. When a Deployment is scaled out, new pods will be created and scheduled to Nodes with available resources. The number of pods will be scaled up to the new target state. Autoscaling pods is also supported by Kubernetes. It is also possible to scale to zero, which will terminate all pods in a given Deployment.

Running many instances of an application needs a method for distributing traffic among them. A built-in load balancer in a `Services` object distributes network traffic across all pods in an exposed Deployment. Endpoints will be used to continuously monitor the operating pods, ensuring that traffic is only directed to those that are available.

We'll use a new YAML file to increase the number of pods in the Deployment in the following example. The replicas are set to 4 in this YAML file, indicating that the Deployment should include four pods:

```
apiVersion: apps/v1
kind: Deployment
metadata:
  name: nginx-deployment
spec:
  selector:
    matchLabels:
      app: nginx
  replicas: 4 # Update the replicas from 2 to 4
  template:
    metadata:
      labels:
        app: nginx
    spec:
      containers:
```

```
- name: nginx
  image: nginx:1.14.2
  ports:
  - containerPort: 80
```

The following command will deploy the preceding updated application deployment:

```
kubectl apply -f deployment-scale.yaml
```

The following command output shows that the command was successfully run and that the Deployment has been configured:

```
ubuntu@controlplane:~$ kubectl apply -f deployment-scale.yaml
deployment.apps/nginx-deployment configured
ubuntu@controlplane:~$
```

Figure 5.29 – Scaling sample application deployment

The following command execution output confirms that there is no error in the deployment, and in the next steps, we can verify that the Deployment has four pods:

```
kubectl get pods -l app=nginx
```

The following command output shows that the command was successfully run and the deployment has created new pods and scheduled to Nodes with available resources:

```
ubuntu@controlplane:~$ kubectl get pods -l app=nginx
NAME                                 READY   STATUS    RESTARTS   AGE
nginx-deployment-9456bbbf9-sgj9b     1/1     Running   0          2m48s
nginx-deployment-9456bbbf9-fmhnb     1/1     Running   0          2m48s
nginx-deployment-9456bbbf9-xsnfj     1/1     Running   0          2m22s
nginx-deployment-9456bbbf9-r6w58     1/1     Running   0          2m22s
ubuntu@controlplane:~$
```

Figure 5.30 – New pods have been created post scaling

Here is a diagram of scaling functionality:

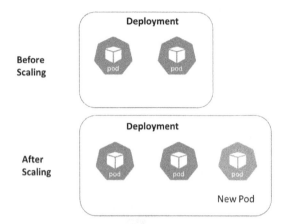

Figure 5.31 – Application deployment scaling

Another option is to use the command line (without editing the YAML file). Let's say we want to increase the number of nginx deployments to five. To do so, run the `kubectl scale` command, as follows:

```
kubectl scale deployments/nginx-deployment --replicas=5
```

Those pods can also be scaled down in the same way they were scaled up. You can alter `replicas:` in the YAML file.

And with the `kubectl` command, you could scale down from 5 to 4, like so:

```
kubectl scale deployments/nginx-deployment --replicas=4
```

To view how the pods are distributed across the Nodes, use the following command to check:

```
kubectl get pods -o wide
```

The following command execution output indicates that there are two pods running on the control plane node and two of them are running on `worker1`:

```
ubuntu@controlplane:~$ kubectl get pods -o wide
NAME                             READY   STATUS    RESTARTS   AGE     IP              NODE
nginx-deployment-9456bbbf9-sgj9b   1/1   Running   0          4m54s   10.1.49.70      controlplane
nginx-deployment-9456bbbf9-fmhnb   1/1   Running   0          4m54s   10.1.235.133    worker1
nginx-deployment-9456bbbf9-xsnfj   1/1   Running   0          4m28s   10.1.49.71      controlplane
nginx-deployment-9456bbbf9-r6w58   1/1   Running   0          4m28s   10.1.235.134    worker1
ubuntu@controlplane:~$
```

Figure 5.32 – Pod distribution across Nodes

In this chapter, we have demonstrated how to get MicroK8s functionality on a Raspberry Pi and have joined multiple Pis to form a production-grade Kubernetes cluster. MicroK8s is being presently employed in a range of contexts, ranging from a single-node installation on a developer's desktop to the support of compute-intensive AI and ML workloads.

MicroK8s is well suited for the edge, IoT, and appliances because of its minimal-resource footprint and support for both **Advanced RISC Machine** (**ARM**) and Intel architectures. On Raspberry Pis, MicroK8s is a common choice. We will be looking at implementation aspects of common edge computing applications in the upcoming chapters. In the next section, we will touch upon some best practices for implementing Kubernetes for your production-grade Kubernetes cluster.

Guidelines on multi-node cluster configuration

In this section, we will go through some best practices for creating a scalable, secure, and highly optimized Kubernetes cluster model.

Cluster-level configuration/settings

You can define levels of abstraction, such as pods and services, with Kubernetes so that you don't have to worry about where your applications are running or whether they have enough resources to run efficiently. However, you must monitor your applications, the containers that run them, and even Kubernetes itself to maintain optimal performance. In this section, we will cover some best practices to follow for setting up and operating your Kubernetes clusters. These are outlined here:

- **Use the latest version**: Kubernetes offers new features, bug patches, and platform upgrades with its regular version updates. You should always utilize the most recent Kubernetes version on your cluster as a rule of thumb.

- When multiple teams are attempting to use the same cluster resources at the same time, namespaces can be used to achieve team-level isolation. Using Namespaces effectively allows you to construct numerous logical cluster divisions, enabling you to allocate different virtual resources to different teams.

- Use smaller container images whenever possible to speed up your builds. Due to a smaller attack surface, smaller images are likewise less vulnerable to attack vectors.

- The **Center for Internet Security** (**CIS**) provides numerous standards and benchmark tests for best practices in code security. They also have a Kubernetes benchmark on their website, which you may download. kube-bench is one of the tools that examine whether Kubernetes is deployed securely using the checks provided in the *CIS Kubernetes Benchmark*.

- Alpha and beta Kubernetes features are still in development and may contain flaws or problems that lead to security flaws. Always weigh the benefits of an alpha or beta feature against the danger posed to your security posture. When in doubt, turn off any features that you don't utilize.

- Use an OpenID Connect authentication mechanism for your Kubernetes cluster and other development tools using **single sign-on (SSO)**, such as Google Identity.

- Review retention and archival strategy for logs—ideally, 30-45 days of historical logs should be retained.

- Logs should be collected from all Nodes, control planes, and auditing:

  a. Nodes (kubelet, container runtime)

  b. Control plane (API server, scheduler, controller manager)

  c. Kubernetes auditing (all requests to the API server)

- Use a log aggregation tool such as the **Amazon Web Services (AWS)** CloudWatch, **Elasticsearch, Fluentd, and Kibana (EFK)** stack, Datadog, Sumo Logic, Sysdig, **Google Cloud Platform (GCP)** Stackdriver, or Azure Monitor.

- Control plane components should be monitored to assist in any discovery of issues/threats within the cluster and reduce latency. It's better to use automatic monitoring tools rather than manually managing alerts.

To summarize, we have covered some of the best practices that need to be followed for setting up and operating your Kubernetes clusters. In the next section, we will look at best practices related to container life cycle management.

Container life cycle management

Kubernetes and the Kubernetes architecture effectively automate the life cycle management of application containers, but they can be difficult to set up and administer. In this section, we will check on best practices and how to implement them in your clusters quickly and easily:

- Containers with no limits might cause resource conflict with other containers and inefficient computational resource consumption. Use `ResourceQuota` and `LimitRange` for restricting resource utilization:

  a. You can use `ResourceQuotas` to set a limit on the total amount of resources consumed by all containers in a Namespace. Other Kubernetes objects, such as the number of pods in the current namespace, can also have quotas imposed.

  b. If you're concerned that someone might use your cluster to produce a large number of ConfigMaps, you can use `LimitRange` to prevent this.

- Use Kubernetes *pod security policies* for enforcing security configurations—for example, to access the host filesystem.

- While there are some circumstances where privileged containers are required, allowing your containers to do so is, in general, a security concern. If there are no specific use cases, disable privileged containers.

- Go for rootless containers. If a user succeeds to break out of a container-based application running as root, they may be able to use the same root user to get access to the host.

- To avoid escalating privileges using `setuid` or `setgid` binaries, run your container with privilege escalation disabled.

- Enable **network policies** so that it establishes firewalls between the pods on your cluster.

- Use tools such as **Open Policy Agent** (**OPA**) to apply policies, such as using only approved base images that can be deployed in your cluster.

To recap, we've gone through some best practices for automating container life cycle management. We'll look at the guidelines for deploying and sharing HA applications with Kubernetes in the next section.

Deploying and sharing HA applications

As you may be aware, deploying a basic application in Kubernetes is a piece of cake. Trying to make your application as available and fault-tolerant as feasible, on the other hand, implies a slew of challenges. In this section, we list some of the guidelines for deploying and sharing HA applications in Kubernetes, as follows:

- All containers should have `readiness probes` set up. The kubelet agent assumes that the application is ready to receive traffic as soon as the container starts if you don't set the readiness probe.

- *Liveness and readiness* probes shouldn't point to the same endpoint because when the application indicates that it is not ready or live, the kubelet agent detaches and deletes the container from the service.

- Running many or more than one instance of your pods ensures that eliminating a single pod will not result in downtime. Also, consider using a Deployment, DaemonSet, ReplicaSet, or StatefulSet to deploy your pod instead of running pods individually.

- *Anti-affinity rules* should be applied to your Deployments so that pods are distributed over all Nodes in your cluster.

- You can set a *pod disruption budget* to safeguard Deployments against unforeseen events that could bring down many pods at the same time. If the final state for that Deployment results in fewer than five pods, Kubernetes will prevent the drain event.

- Use the resources property of `containerSpec` to specify resource constraints that limit how much CPU and memory your containers can consume. These settings are taken into consideration by the scheduler to determine which node is most suited for the current pod.

- All resources should have technical, business, and security labels defined. They should be applied to all resources in your cluster, including pods, services, Ingress manifests, and endpoints.

- Containers should not store any state in their local filesystem. Any persistent data should instead be saved in a central location outside of the pods—for instance, in a clustered PersistentVolume, or—even better—in a storage system outside of your cluster.

ConfigMaps should be used to manage all configurations outside of the application code. ConfigMaps should only be used to save non-sensitive settings. For sensitive information, use a secret resource (such as credentials). Instead of being passed in as environment variables, secret resources should be mounted as volumes in containers. To summarize, we have looked at best practices for creating a scalable, secure, and highly optimized Kubernetes cluster model.

Summary

In this chapter, we learned how to set up a MicroK8s Raspberry Pi multi-node cluster, deployed a sample application, and executed rolling updates on the deployed application. We also discovered ways to scale the deployed application. We also found that, while Kubernetes allows us to define levels of abstraction such as pods and services to help with application deployments, we must monitor the applications, containers, clusters, and Kubernetes itself to ensure optimal performance. In this context, we learned about several recommended practices for building a scalable, secure, and highly optimized Kubernetes cluster model.

In the next chapter, we will look at how to configure container network connectivity for your Kubernetes cluster.

6
Configuring Connectivity for Containers

We learned how to set up a MicroK8s Raspberry Pi multi-node cluster, deploy a sample application, and perform rolling updates on the deployed application in the previous chapter. We also figured out how to scale the deployed application. We also learned about a few best practices for designing a Kubernetes cluster that is scalable, secure, and highly optimized. In this and the following chapters, we'll continue to implement various use cases of common edge-computing applications using MicroK8s. Kubernetes provides several ways for exposing Services to the outside world.

In this chapter, we'll continue with our next use case, which is about container network connectivity on MicroK8s. Each Pod in the Kubernetes network model is assigned its own **Internet Protocol** (**IP**) address by default. As a result, you won't have to explicitly link or network Pods together, and you shouldn't have to bother with mapping container ports to host ports, and so on.

Kubernetes allows you to describe declaratively how your applications are deployed, how they communicate with one another and with the Kubernetes control plane, and how clients can access them.

Kubernetes, as a highly modular open source project, allows for a great level of network implementation adaptability. The Kubernetes ecosystem has spawned a slew of projects aimed at making container communication simple, consistent, and safe. One project that enables plugin-based features to ease networking in Kubernetes is **Container Network Interface** (**CNI**). The major goal of CNI is to give administrators enough control to monitor traffic while decreasing the time it takes to manually configure network configurations.

The following fundamental criteria are imposed by Kubernetes on any networking implementation:

- Without **network address translation** (**NAT**), Pods on a node can communicate with Pods on all other nodes.
- A node's agents (such as system daemons and `kubelet`) can communicate with all the node's Pods.
- Without NAT, Pods in a node's host network can communicate with Pods on all other nodes.

CNI allows a Kubernetes provider to develop unique networking models that seek to deliver a consistent and dependable network across all your Pods. CNI plugins provide namespace isolation, traffic, and IP filtering, which Kubernetes does not provide by default. Let's say a programmer wishes to use these advanced network functionalities. In such situations, they must utilize the CNI plugin in conjunction with CNI to facilitate network construction and administration.

There are a variety of CNI plugins on the market. In this chapter, we will look at some of the popular options—such as Flannel, Calico, and Cilium—and we're going to cover the following main topics:

- CNI overview
- Configuring Calico
- Configuring Cilium
- Configuring Flannel
- Guidelines on choosing a CNI provider

CNI overview

Before diving into a CNI overview, let's understand how networking is handled within a Kubernetes cluster.

When Kubernetes schedules a Pod to execute on a node, the node's Linux kernel generates a network namespace for the Pod. This network namespace establishes a **virtual network interface** (**VIF**) between the node's physical network interface—such as eth0—and the Pod, allowing packets to flow to and from the Pod. The related VIF in the root network namespace of the node connects to a Linux bridge, allowing communication between Pods on the same node. A Pod can also use the same VIF to send packets outside of the node.

From a range of addresses reserved for Pods on the node, Kubernetes assigns an IP address (Pod IP address) to the VIF in the Pod's network namespace. This address range is a subset of the cluster's IP address range for Pods, which you can specify when you build a cluster.

The network namespace used by a container running in a Pod is the Pod's network namespace. The Pod seems to be a physical machine with one network interface from the perspective of the container. This network interface is shared by all containers in the Pod. The localhost of each container is connected to the node's physical network interface, such as eth0, via the Pod. Each Pod has unfiltered access to all other Pods operating on all cluster nodes by default, but you can restrict access among Pods.

The following diagram depicts a single node running two Pods and the network traffic between the Pods:

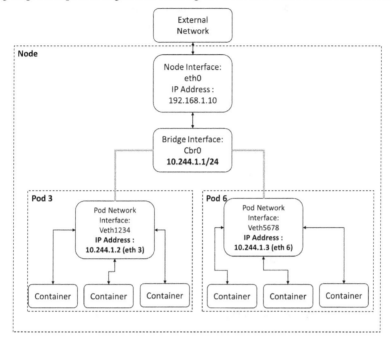

Figure 6.1 – Kubernetes network model: Flow of traffic between Pods

Communication flow from Pod 3 to Pod 6

Let's look at the communication flow from Pod3 to Pod6 which is housed in a single node:

1. A packet leaves from Pod 3 through the eth3 interface and reaches the cbr0 bridge interface through the veth1234 virtual interface.
2. The packet leaves veth1234 and reaches cbr0, looking for the address of Pod 6.
3. The packet leaves cbr0 and is redirected to veth5678.
4. The packet leaves cbr0 through veth5678 and reaches the Pod 6 network through the eth6 interface.

On a regular basis, Kubernetes destroys and rebuilds Pods. As a result, Services that have a stable IP address and enable load balancing among a set of Pods must be used. The kube-proxy component residing in the node takes care of communication between Pods and Services.

The flow of traffic from a client Pod 3 to a server Pod 6 on a separate node is depicted in the following diagram. The Kubernetes **application programming interface** (**API**) server keeps track of the application's Pods. This list is used by the kube-proxy agent process on each node to configure an iptables rule that directs traffic to the proper Pod:

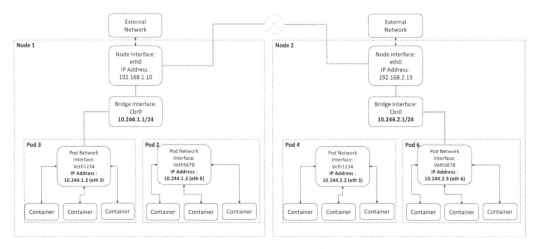

Figure 6.2 – Kubernetes network model: Flow of traffic between Pods on different nodes

Communication flow from Pod 3 to Pod 6 on different nodes

Let's look at the communication flow from Pod3 to Pod6 which is housed in different nodes:

1. A packet leaves from Pod 3 through the eth3 interface and reaches the cbr0 bridge interface through the veth1234 virtual interface.
2. The packet leaves veth1234 and reaches cbr0, looking for the address of Pod 6.
3. The packet leaves cbr0 and is redirected to eth0.
4. The packet then leaves eth0 from node 1 and reaches the gateway.
5. The packet leaves the gateway and reaches the eth0 interface on node 2.
6. The packet leaves eth0 and reaches cbr0, looking for the address of Pod 6.
7. The packet leaves cbr0 through veth5678 and reaches the Pod 6 network through the eth6 interface.

Now that we are clear on how the traffic flow is routed in a Kubernetes network model, we can now focus on CNI concepts.

CNI is a network framework that uses a set of standards and modules to enable the dynamic setup of networking resources. The plugin's specification details the interface for configuring the network, provisioning IP addresses, and maintaining multi-host communication.

CNI effortlessly connects with the `kubelet` agent in the Kubernetes context to allow automatic network configuration between Pods, utilizing either an underlay or an overlay network. Let's look at this in more detail here:

- **Overlay mode**—A container in **Overlay** mode is independent of the host's IP address range. Tunnels are established between hosts during cross-host communication, and all packets in the container **Classless Inter-Domain Routing (CIDR)** block are encapsulated (using a virtual interface such as **Virtual eXtensible Local Area Network**, or **VXLAN**) as packets exchanged between hosts in the underlying physical network. This mode eliminates the underlying network's dependency, and you can see an overview of it in the following diagram:

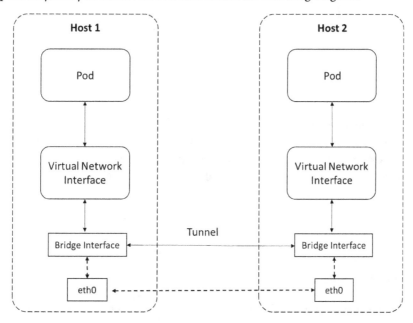

Figure 6.3 – Overlay mode

- **Underlay mode**—Containers and hosts are located at the same network layer and share the same position in **Underlay** mode. Container network interconnection is determined by the underlying network (physical level of the networking layer), which consists of routers and switches. As a result, the underlying capabilities are heavily reliant on this mode. You can see an overview of this in the following diagram:

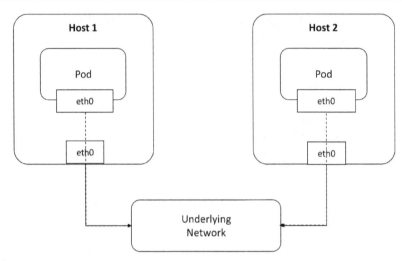

Figure 6.4 – Underlay mode

Once the network configuration type is defined, the runtime creates a network for containers to join and uses the CNI plugin to add the interface to the container namespace and use the **IP Address Management (IPAM)** plugin to allocate the linked subnetwork and routes. In addition to Kubernetes networking, CNI also supports a **software-defined networking (SDN)** approach to offer unified container communication across a cluster.

Now that we are clear on CNI concepts, we will delve into the steps of configuring Calico CNI plugin to network across a cluster.

Configuring Calico

Calico is the most popular open source CNI plugin for the Kubernetes environment. **Tigera** maintains Calico, which is intended for use in contexts where network performance, flexibility, and power are crucial. It has strong network administration security capabilities, as well as a comprehensive view of host and Pod connectivity.

It can be easily deployed as a `DaemonSet` on each node in a regular Kubernetes cluster. For managing numerous networking activities, each node in a cluster would have three Calico components installed: `Felix`, `BIRD`, and `confd`. Node routing is handled by `Felix`, a Calico agent, while `BIRD` and `confd` manage routing configuration changes.

Calico uses the **Border Gateway Protocol (BGP)** routing protocol instead of an overlay network to route messages between nodes. IP-IN-IP or VXLAN, which may encapsulate packets delivered across subnets such as an overlay network, provide an overlay networking mode. It employs an unencapsulated IP network fabric, which reduces the need to encapsulate packets, resulting in improved network performance for Kubernetes workloads.

WireGuard, which establishes and manages tunnels between nodes to ensure secure communication, encrypts in-cluster Pod communications. It makes tracing and debugging a lot easier than other tools because it doesn't use wrappers to manipulate packets. Developers and administrators can quickly analyze packet behavior and take advantage of complex network features such as policy management and **access control lists** (**ACLs**).

Calico's network policies implement deny/match rules that may be applied to Pods using manifests to assign ingress policies. To monitor Pod traffic, boost security, and govern Kubernetes workloads, users can build globally scoped policies and interface with an Istio Service mesh.

In the following steps, we will demonstrate how Calico can secure your Kubernetes cluster with a basic example of the Kubernetes NetworkPolicy API. NetworkPolicies are application-centric constructs that allow you to declare how Pods can communicate across the network with various network entities.

The entities with which a Pod can communicate are identified using a combination of the three **identifiers** (**IDs**), shown next:

- Other Pods that are permissible (exception: a Pod cannot block access to itself)
- Namespaces that are allowed
- IP blocks that are allowed (exception: traffic to and from the node where a Pod is running is always allowed, regardless of the IP address of the Pod or the node)

Now that we are clear on the IDs, we will delve into the steps of configuring Calico CNI plugin to network across a cluster. The following diagram depicts our Raspberry Pi cluster setup:

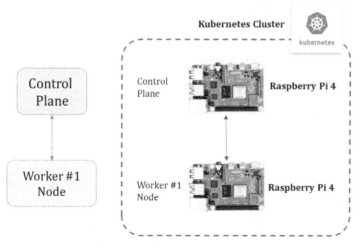

Figure 6.5 – Raspberry Pi cluster setup

Now that we know what we want to do, let's look at the requirements.

Requirements

Before you begin, here are the prerequisites that are needed for building a Raspberry Pi Kubernetes cluster and for the configuration of the CNI:

- A microSD card (4 **gigabytes** (**GB**) minimum; 8 GB recommended)
- A computer with a microSD card drive
- A Raspberry Pi 2, 3, or 4 (one or more)
- A micro-USB power cable (USB-C for the Pi 4)
- A Wi-Fi network or an Ethernet cable with an internet connection
- (Optional) A monitor with a **High-Definition Multimedia Interface** (**HDMI**) interface
- (Optional) An HDMI cable for the Pi 2 and 3 and a micro-HDMI cable for the Pi 4
- (Optional) A **Universal Serial Bus** (**USB**) keyboard

Now that we've established the requirements, we'll go on to the step-by-step instructions on how to complete the process.

Step 1 – Creating a MicroK8s Raspberry Pi cluster

Please follow the steps that we covered in *Chapter 5, Creating and Implementing Updates on Multi-Node Raspberry Pi Kubernetes Clusters*, to create a MicroK8s Raspberry Pi cluster. Here is a quick refresher:

- *Step 1*: Installing **operating system** (**OS**) image to SD card
- *Step 1a*: Configuring Wi-Fi access settings
- *Step 1b*: Configuring remote access settings
- *Step 1c*: Configuring control group settings
- *Step 1d*: Configuring hostname
- *Step 2*: Installing and configuring MicroK8s
- *Step 3*: Adding a worker node

A fully functional multi-node Kubernetes cluster would look like the one shown next. To summarize, we have installed MicroK8s on the Raspberry Pi boards and joined multiple deployments to form the cluster. We have also added nodes to the cluster:

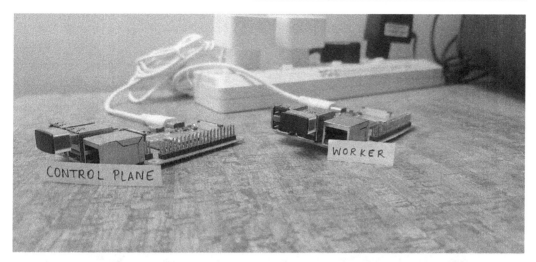

Figure 6.6 – Fully functional MicroK8s Kubernetes cluster

We can go to the next step of enabling the Calico add-on now that we have a fully functional cluster.

Step 2 – Enabling the Calico CNI add-on

By default, Calico is enabled if a cluster add-on is enabled. We can verify whether it's enabled by using the following command:

```
kubectl get pods - A | grep calico
```

The following command execution output indicates Calico is enabled and its Pods are running:

```
ubuntu@controlplane:~$ kubectl get pods -A |grep calico
kube-system        calico-kube-controllers-548d5485bf-tfnhc      1/1        Running
kube-system        calico-node-n6xkg                             1/1        Running
ubuntu@controlplane:~$
```

Figure 6.7 – Validating Calico Pods are running

Now that we have Calico CNI running, let's create a sample nginx deployment for us to test the network isolation in the next step. By default, a Pod is not isolated for egress and ingress—that is, all outbound and inbound connections are allowed.

Step 3 – Deploying a sample containerized application

Use the following command to create a sample nginx deployment:

```
kubectl create deployment nginx --image=nginx
```

The following command execution output indicates that there is no error in the deployment, and in the next steps, we can expose the `nginx` deployment that we created:

```
ubuntu@controlplane:~$ kubectl create deployment nginx --image=nginx
deployment.apps/nginx created
ubuntu@controlplane:~$
```

Figure 6.8 – Sample application deployment

Use the following command to expose the `nginx` deployment so that it can be accessed from other Pods:

```
kubectl expose deployment nginx --port=80
```

The following command execution output confirms that expose deployment has succeeded:

```
ubuntu@controlplane:~$ kubectl expose deployment nginx --port=80
service/nginx exposed
ubuntu@controlplane:~$
```

Figure 6.9 – Exposing the sample application

Use the following command to see whether the Service has been exposed:

```
kubectl get svc
```

The following command execution output shows that the Service is exposed, and a cluster IP has been assigned. Using the cluster IP and port, we can access the Service from other Pods. Recall from *Chapter 6, Setting up MetalLB and Ingress for Load Balancing*, that an external IP would not have been allocated because an external load balancer such as `MetalLB` must be enabled for this:

```
ubuntu@controlplane:~$ kubectl get svc | grep nginx
nginx              ClusterIP      10.152.183.117    <none>        80/TCP          37s
ubuntu@controlplane:~$
```

Figure 6.10 – Cluster IP for the Service is allocated

We'll establish a new Pod to access the Service now that the Services have been exposed. Use the following command to create a new Pod and open up a shell session inside the Pod:

```
kubectl run access --rm -ti --image busybox /bin/sh
```

The following command execution output confirms that the `run` command has succeeded and a shell session has opened up inside the `access` Pod:

```
ubuntu@controlplane:~$ kubectl run access --rm -ti --image busybox /bin/sh
If you don't see a command prompt, try pressing enter.
/ #
```

Figure 6.11 – Shell session for the access Pod

Use the following command to access the nginx Service from the access Pod:

```
wget -q nginx -O -
```

Great! The nginx Service is accessible from the access Pod, as we can see here:

```
/ # wget -q nginx -O -
<!DOCTYPE html>
<html>
<head>
<title>Welcome to nginx!</title>
<style>
html { color-scheme: light dark; }
body { width: 35em; margin: 0 auto;
font-family: Tahoma, Verdana, Arial, sans-serif; }
</style>
</head>
<body>
<h1>Welcome to nginx!</h1>
<p>If you see this page, the nginx web server is successfully installed and
working. Further configuration is required.</p>

<p>For online documentation and support please refer to
<a href="http://nginx.org/">nginx.org</a>.<br/>
Commercial support is available at
<a href="http://nginx.com/">nginx.com</a>.</p>

<p><em>Thank you for using nginx.</em></p>
</body>
</html>
/ #
```

Figure 6.12 – nginx response

To summarize, we've set up a test nginx application and exposed and tested the Service from the access Pod. Isolation will be applied in the next step by using NetworkPolicy.

Step 4 – Applying isolation by using NetworkPolicy

Let's create a NetworkPolicy for all Pods in the default namespace that implements a default deny behavior, as follows:

```
kind: NetworkPolicy
apiVersion: networking.k8s.io/v1
```

```
metadata:
  name: default-deny
spec:
  podSelector:
    matchLabels: {}
```

From the preceding code, we can note that `podSelector` is included in each NetworkPolicy, which selects the grouping of Pods to which the policy applies. In the preceding policy, an empty `podSelector` indicates that it applies to all Pods in the namespace.

Use the following command to create isolation using NetworkPolicy:

`kubectl apply -f calico-policy.yaml`

The following command execution output confirms that there is no error in the deployment and Calico will then block all connections to Pods in this namespace:

```
ubuntu@controlplane:~$ kubectl apply -f calico-policy.yaml
networkpolicy.networking.k8s.io/default-deny created
ubuntu@controlplane:~$ 
```

Figure 6.13 – NetworkPolicy created

To test access to the `nginx` Service, run the following command from within the BusyBox `access` Pod:

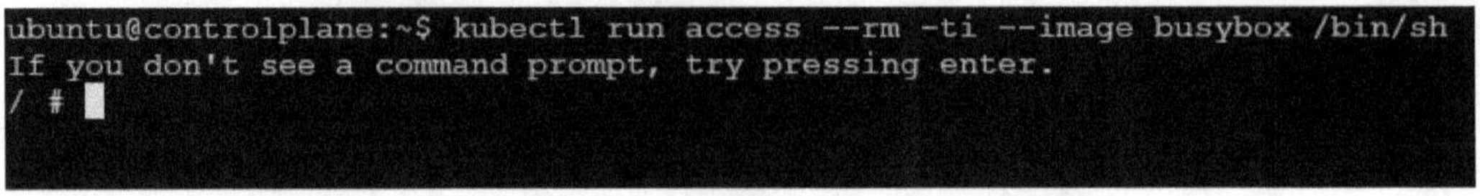

```
ubuntu@controlplane:~$ kubectl run access --rm -ti --image busybox /bin/sh
If you don't see a command prompt, try pressing enter.
/ # 
```

Figure 6.14 – Testing access to the nginx Service from the access Pod

Use the same `wget` command to access the `nginx` Service from the `access` Pod, as follows:

`wget -q nginx -O -`

The following command output confirms that the `nginx` Service is not accessible, so let's try with a timed-out setting in the next step, as follows:

```
wget: download timed out
/ # wget -q nginx -O -

```

Figure 6.15 – Using the wget command to access nginx

The following command output confirms the request timed out after 5 seconds:

```
/ # wget -q --timeout=5 nginx -O -
wget: download timed out
```

Figure 6.16 – Request timed out after 5 seconds

Now that we've tested the isolation using the deny rule, it's time to provide access and test the incoming connections.

Step 5 – Enabling access

Let's modify the same NetworkPolicy to grant access to the nginx Service. Incoming connections from our access Pod only will be allowed, but not from anywhere else. The code is illustrated in the following snippet:

```
kind: NetworkPolicy
apiVersion: networking.k8s.io/v1
metadata:
  name: access-nginx
spec:
  podSelector:
    matchLabels:
      app: nginx
  ingress:
    - from:
      - podSelector:
          matchLabels:
            run: access
```

From the preceding code, we can note the following:

- podSelector selects Pods with matching labels of type app: nginx.

- The ingress rule allows traffic if it matches the from section.

Use the following command to apply the modified policy:

```
kubectl apply -f calico-policy.yaml
```

The following command execution output confirms that there is no error in the deployment. Traffic from Pods with the `run: access` label to Pods with the `app: nginx` label is allowed by the NetworkPolicy. Labels are created automatically by `kubectl` and are based on the resource name:

```
ubuntu@controlplane:~$ kubectl apply -f calico-policy.yaml
networkpolicy.networking.k8s.io/access-nginx created
ubuntu@controlplane:~$
```

Figure 6.17 – Deployment of the modified policy

Use the `wget` command to access the `nginx` Service from the `access` Pod, as follows:

```
wget -q nginx -O -
```

The following output of the preceding command confirms that we can access the `nginx` Service from the `access` Pod:

```
/ # wget -q nginx -O -
<!DOCTYPE html>
<html>
<head>
<title>Welcome to nginx!</title>
<style>
html { color-scheme: light dark; }
body { width: 35em; margin: 0 auto;
font-family: Tahoma, Verdana, Arial, sans-serif; }
</style>
</head>
<body>
<h1>Welcome to nginx!</h1>
<p>If you see this page, the nginx web server is successfully installed and
working. Further configuration is required.</p>

<p>For online documentation and support please refer to
<a href="http://nginx.org/">nginx.org</a>.<br/>
Commercial support is available at
<a href="http://nginx.com/">nginx.com</a>.</p>

<p><em>Thank you for using nginx.</em></p>
</body>
</html>
/ #
```

Figure 6.18 – Testing access using the wget command

To reconfirm, let's create a Pod without the `run: access` label using the following command and test whether it's working correctly:

```
kubectl run access1 --rm -ti --image busybox /bin/sh
```

As shown in the following command execution output, this should start a shell session inside the
access1 Pod:

```
ubuntu@controlplane:~$ kubectl run access1 --rm -ti --image busybox /bin/sh
If you don't see a command prompt, try pressing enter.
/ #
```

Figure 6.19 – Shell session inside the access1 Pod

To test access to the nginx Service, run the wget command from within the BusyBox access1
Pod, as follows:

```
wget -q nginx -O -
```

The request should time out since the NetworkPolicy will allow only access from a Pod with a
run: access label, as illustrated here:

```
ubuntu@Master:~$ kubectl run access1 --rm -ti --image busybox /bin/sh
If you don't see a command prompt, try pressing enter.
/ #  wget -q nginx -O -
^C
/ #  wget -q --timeout=5 nginx -O -
wget: download timed out
/ #
```

Figure 6.20 – Request timed out

This was just a quick demonstration of the Kubernetes NetworkPolicy API and how Calico can help
you secure your Kubernetes cluster. For more information about Kubernetes network policy, refer to
the following link: https://kubernetes.io/docs/concepts/services-networking/
network-policies/.

Calico's powerful network policy framework makes it simple to restrict communication so that only the
traffic you want flows. Furthermore, with built-in WireGuard encryption functionality, safeguarding
your Pod-to-Pod traffic across the network has never been easier.

Calico's policy engine can enforce the same policy model at the host networking layer, safeguarding your
infrastructure from compromised workloads and your workloads from compromised infrastructure.

Calico is a great option for consumers who desire complete control over their network components.
It is also compatible with a variety of Kubernetes platforms and provides commercial support via
Calico Enterprise.

The highly scalable Cilium CNI solution created by Linux kernel developers will be discussed in the
following section.

Configuring Cilium

Cilium uses **extended Berkeley Packet Filter (eBPF)** filtering technology to ensure network connectivity across Kubernetes Services by adding high-level application rules. It manages operations and converts network definitions to eBPF applications as a `cilium-agent` daemon on each node of the Kubernetes cluster. Pods communicate with one another using an overlay network or a routing mechanism; for instance, both IPv4 and IPv6 addresses are supported. VXLAN tunneling is used for packet encapsulation in overlay networks, while native routing is done via the unencapsulated BGP protocol.

The eBPF Linux kernel feature allows for the dynamic insertion of sophisticated security visibility and control logic within Linux itself, as shown in the following diagram. Cilium security policies can be applied and modified without requiring any changes to the application code or container configuration because eBPF operates inside the Linux kernel. It also has **HyperText Transfer Protocol (HTTP)** request filters that support Kubernetes Network Policies. Both ingress and egress enforcements are available, and the policy configuration can be expressed in **YAML Ain't Markup Language (YAML)** or **JavaScript Object Notation (JSON)** format. While integrating policies with service meshes such as Istio, administrators can approve or reject requests based on the request method or path header:

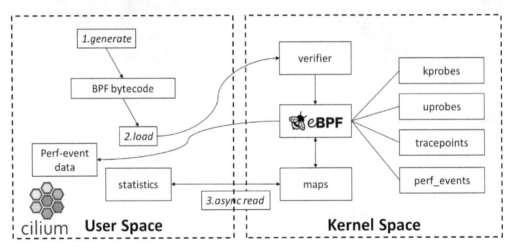

- Network Policy
- Services & Load Balancing
- Bandwidth Management
- Flow & Policy Logging
- Metrics

Figure 6.21 – Cilium: eBPF-based networking, observability, and security

More details about eBPF technology can be found here: `https://ebpf.io/`.

Cilium may be utilized across several Kubernetes clusters and provides multi-CNI functionality, a high level of inspection, and Pod-to-Pod interaction. Packet inspection and application protocol packets are managed by its network- and application-layer awareness.

In the next steps, we will be using Kubernetes' `NetworkPolicy`, `CiliumNetworkPolicy`, and `CiliumClusterwideNetworkPolicy` resources to apply policies to our cluster. Kubernetes will automatically distribute the policies to all agents.

Since Cilium isn't available for `arm64` architecture, I'll be using an Ubuntu **virtual machine** (**VM**) for this section. The instructions for setting up a MicroK8s cluster are the same as in *Chapter 5, Creating and Implementing Updates on Multi-node Raspberry Pi Kubernetes Clusters*.

Step 1 – Enabling the Cilium add-on

Use the following command to enable the Cilium add-on:

```
microk8s enable cilium
```

The following command execution output indicates the Cilium add-on has been enabled successfully:

```
ubuntu@microk8s2-worker1:~$ microk8s enable cilium
Enabling Helm 3
Fetching helm version v3.5.0.
  % Total    % Received % Xferd  Average Speed   Time    Time     Time  Current
                                 Dload  Upload   Total   Spent    Left  Speed
100 11.7M  100 11.7M    0     0  7403k      0  0:00:01  0:00:01 --:--:-- 7403k
Helm 3 is enabled
Restarting kube-apiserver
Restarting kubelet
Enabling Cilium
Fetching cilium version v1.10.
  % Total    % Received % Xferd  Average Speed   Time    Time     Time  Current
                                 Dload  Upload   Total   Spent    Left  Speed
100   131    0   131    0     0   433      0 --:--:-- --:--:-- --:--:--   436
100 32.2M    0 32.2M    0     0  6773k      0 --:--:--  0:00:04 --:--:-- 7318k
Deploying /var/snap/microk8s/2948/actions/cilium.yaml. This may take several minutes.
serviceaccount/cilium created
serviceaccount/cilium-operator created
configmap/cilium-config created
clusterrole.rbac.authorization.k8s.io/cilium created
clusterrole.rbac.authorization.k8s.io/cilium-operator created
clusterrolebinding.rbac.authorization.k8s.io/cilium created
clusterrolebinding.rbac.authorization.k8s.io/cilium-operator created
Warning: spec.template.spec.affinity.nodeAffinity.requiredDuringSchedulingIgnoredDuringE
[0].key: beta.kubernetes.io/os is deprecated since v1.14; use "kubernetes.io/os" instead
Warning: spec.template.metadata.annotations[scheduler.alpha.kubernetes.io/critical-pod]:
assName" field instead
daemonset.apps/cilium created
deployment.apps/cilium-operator created
Waiting for daemon set spec update to be observed...
```

Figure 6.22 – Enabling Cilium add-on

It will take some time to finish activating the add-on, but the following command execution output shows that Cilium has been successfully enabled:

```
clusterrole.rbac.authorization.k8s.io "calico-kube-controllers" deleted
clusterrolebinding.rbac.authorization.k8s.io "calico-kube-controllers" deleted
clusterrole.rbac.authorization.k8s.io "calico-node" deleted
clusterrolebinding.rbac.authorization.k8s.io "calico-node" deleted
daemonset.apps "calico-node" deleted
serviceaccount "calico-node" deleted
deployment.apps "calico-kube-controllers" deleted
serviceaccount "calico-kube-controllers" deleted
Warning: policy/v1beta1 PodDisruptionBudget is deprecated in v1.21+, unavailable in v1.25+; use
poddisruptionbudget.policy "calico-kube-controllers" deleted
Cilium is enabled
ubuntu@microk8s2-worker1:~$
```

Figure 6.23 – Cilium add-on enabled

Cilium is now configured! We can now use the microk8s.cilium **command-line interface** (CLI) to check the status of the Cilium configuration. The following command execution output indicates Cilium CNI has been enabled successfully and the controller status is healthy:

```
ubuntu@microk8s:~$microk8s.cilium status
KVStore:                Ok    Disabled
Kubernetes:             Ok    1.23+ (v1.23.3-2+d441060727c463) [linux/amd64]
Kubernetes APIs:        ["cilium/v2::CiliumClusterwideNetworkPolicy", "cilium/v2::CiliumEndpoint", "cilium/v2::CiliumNetworkPolicy", "
cilium/v2::CiliumNode", "core/v1::Namespace", "core/v1::Node", "core/v1::Pods", "core/v1::Service", "discovery/v1::EndpointSlice", "ne
tworking.k8s.io/v1::NetworkPolicy"]
KubeProxyReplacement:   Disabled
Cilium:                 Ok    1.10.7 (v1.10.7-3e77756)
NodeMonitor:            Listening for events on 1 CPUs with 64x4096 of shared memory
Cilium health daemon:   Ok
IPAM:                   IPv4: 2/254 allocated from 10.0.0.0/24,
BandwidthManager:       Disabled
Host Routing:           Legacy
Masquerading:           IPTables [IPv4: Enabled, IPv6: Disabled]
Controller Status:      18/18 healthy
Proxy Status:           OK, ip 10.0.0.205, 0 redirects active on ports 10000-20000
Hubble:                 Ok    Current/Max Flows: 100/4095 (2.44%), Flows/s: 0.15   Metrics: Disabled
Encryption:             Disabled
Cluster health:         1/1 reachable   (2022-02-17T13:13:35Z)
ubuntu@microk8s:~$
```

Figure 6.24 – Cilium CNI

Now that Cilium CNI has been successfully activated and the controller status has been verified as healthy, the next step is to enable the **Domain Name System** (DNS) add-on.

Step 2 – Enabling the DNS add-on

Since we need address resolution services, we are going to enable the DNS add-on as well. The following command execution output indicates DNS has been enabled successfully:

```
ubuntu@microk8s:~$microk8s enable dns
Enabling DNS
Applying manifest
serviceaccount/coredns created
configmap/coredns created
deployment.apps/coredns created
service/kube-dns created
clusterrole.rbac.authorization.k8s.io/coredns created
clusterrolebinding.rbac.authorization.k8s.io/coredns created
Restarting kubelet
DNS is enabled
```

Figure 6.25 – Enabling DNS add-on

Now that we have Cilium CNI running, let's create a sample `nginx` deployment for us to test the network isolation in the next step.

Step 3 – Deploying a sample containerized application

Use the following command to create a sample `nginx` deployment:

```
kubectl create deployment nginx-cilium --image=nginx
```

The following command execution output indicates that there is no error in the deployment, and in the next steps, we can expose the `nginx-cilium` deployment we just created:

```
ubuntu@microk8s:~$kubectl create deployment nginx-cilium --image=nginx
deployment.apps/nginx-cilium created
ubuntu@microk8s:~$
```

Figure 6.26 – Sample application deployment

Use the `kubectl expose` command to expose the `nginx` deployment so that it can be accessed from other Pods, as follows:

```
kubectl expose deployment nginx-cilium --port=80
```

The following command execution output confirms that the expose deployment has succeeded:

```
ubuntu@microk8s:~$kubectl expose deployment nginx-cilium --port=80
service/nginx-cilium exposed
ubuntu@microk8s:~$
```

Figure 6.27 – Exposing the sample application

Now that the Services have been exposed, we'll create a new Pod to access them. To build a new Pod and open a shell session inside it, run the following command:

```
kubectl run access --rm -ti --image busybox /bin/sh
```

The following command execution output confirms that the `run` command has succeeded and a shell session has opened up inside the `access` Pod. We can now confirm that the `nginx-cilium` Service can be accessed from the `access` Pod:

```
ubuntu@microk8s:~$kubectl run access --rm -ti --image busybox /bin/sh
If you don't see a command prompt, try pressing enter.
/ # wget -q nginx-cilium -O -
<!DOCTYPE html>
<html>
<head>
<title>Welcome to nginx!</title>
<style>
html { color-scheme: light dark; }
body { width: 35em; margin: 0 auto;
font-family: Tahoma, Verdana, Arial, sans-serif; }
</style>
</head>
<body>
<h1>Welcome to nginx!</h1>
<p>If you see this page, the nginx web server is successfully installed and
working. Further configuration is required.</p>

<p>For online documentation and support please refer to
<a href="http://nginx.org/">nginx.org</a>.<br/>
Commercial support is available at
<a href="http://nginx.com/">nginx.com</a>.</p>

<p><em>Thank you for using nginx.</em></p>
</body>
</html>
/ #
```

Figure 6.28 – nginx response

To summarize, we created a test `nginx-cilium` application and exposed and tested it from the `access` Pod. In the next stage, NetworkPolicy will be used to test the isolation.

Step 4 – Applying isolation by using NetworkPolicy

Let's create a NetworkPolicy for all Pods in the default namespace that implements a default deny behavior, as follows:

```
kind: NetworkPolicy
apiVersion: networking.k8s.io/v1
metadata:
  name: cilium-deny
spec:
  podSelector:
    matchLabels: {}
```

From the preceding code, we can note that podSelector is included in each NetworkPolicy, which selects the grouping of Pods to which the policy applies. In the preceding policy, an empty podSelector indicates that it applies to all Pods in the namespace.

Use the kubectl apply command to create isolation using NetworkPolicy, as follows:

```
kubectl apply -f cilium-policy.yaml
```

The following command execution output confirms that there is no error in the deployment, and Cilium will then block all connections to Pods in this namespace:

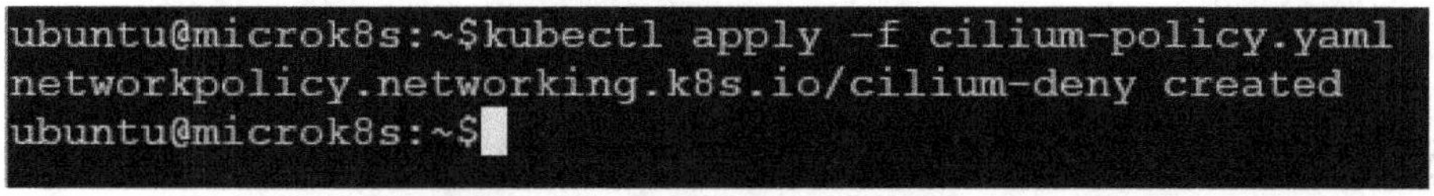

```
ubuntu@microk8s:~$kubectl apply -f cilium-policy.yaml
networkpolicy.networking.k8s.io/cilium-deny created
ubuntu@microk8s:~$
```

Figure 6.29 – NetworkPolicy created

We can also verify the same using the MicroK8s Cilium CLI as well. The following command execution output confirms that a policy has been created:

```
ubuntu@microk8s:~$microk8s cilium policy get
[
  {
    "endpointSelector": {
      "matchLabels": {
        "k8s:io.kubernetes.pod.namespace": "default"
      }
    },
    "ingress": [
      {}
    ],
    "labels": [
      {
        "key": "io.cilium.k8s.policy.derived-from",
        "value": "NetworkPolicy",
        "source": "k8s"
      },
      {
        "key": "io.cilium.k8s.policy.name",
        "value": "cilium-deny",
        "source": "k8s"
      },
      {
        "key": "io.cilium.k8s.policy.namespace",
        "value": "default",
        "source": "k8s"
      },
      {
        "key": "io.cilium.k8s.policy.uid",
        "value": "46c9e099-1fc0-49b8-9441-32471f47624d",
        "source": "k8s"
      }
    ]
  }
]
Revision: 2
ubuntu@microk8s:~$
```

Figure 6.30 – Cilium CLI shows a policy has been created

To test access to the nginx-cilium Service, run the wget command from within the BusyBox access Pod, as follows:

```
wget -q --timeout=5 nginx-cilium -O -
```

The following command output confirms that the nginx-cilium Service is not accessible, and the request timed out after 5 seconds:

```
/ # wget -q --timeout=5 nginx-cilium -O -
wget: download timed out
/ #
```

Figure 6.31 – Testing access to the nginx-cilium Service from the access Pod

Now that we've tested isolation using the deny rule, it's time to provide access and test incoming connections as well.

Step 5 – Enabling access

Let's modify the same NetworkPolicy to grant access to the `nginx-cilium` Service. Incoming connections from our access Pod only will be allowed, but not from anywhere else. The code is illustrated in the following snippet:

```
kind: NetworkPolicy
apiVersion: networking.k8s.io/v1
metadata:
  name: access-nginx
spec:
  podSelector:
    matchLabels:
      app: nginx
  ingress:
    - from:
      - podSelector:
          matchLabels:
            run: access
```

From the preceding code, we can note the following:

- `podSelector` selects Pods with matching labels of type `app: nginx`.

- The `ingress` rule allows traffic if it matches the `from` section.

Use the following command to apply the modified isolation using NetworkPolicy:

```
kubectl apply -f cilium-policy.yaml
```

The following command execution output confirms that there is no error in the deployment. The NetworkPolicy allows traffic from Pods with the `run: access` label to flow to Pods with the `app: nginx.` label. Labels are generated by `kubectl` automatically and are based on the resource name:

```
ubuntu@microk8s:~$kubectl apply -f cilium-policy.yaml
networkpolicy.networking.k8s.io/cilium-deny configured
ubuntu@microk8s:~$
```

Figure 6.32 – Deployment of the modified policy

We can also verify the same using the MicroK8s Cilium CLI as well. The following command execution output confirms that the policy has been updated:

```
ubuntu@microk8s:~$microk8s cilium policy get
[
  {
    "endpointSelector": {
      "matchLabels": {
        "k8s:app": "nginx-cilium",
        "k8s:io.kubernetes.pod.namespace": "default"
      }
    },
    "ingress": [
      {
        "fromEndpoints": [
          {
            "matchLabels": {
              "k8s:io.kubernetes.pod.namespace": "default",
              "k8s:run": "access"
            }
          }
        ]
      }
    ],
    "labels": [
      {
        "key": "io.cilium.k8s.policy.derived-from",
        "value": "NetworkPolicy",
        "source": "k8s"
      },
      {
        "key": "io.cilium.k8s.policy.name",
        "value": "cilium-deny",
        "source": "k8s"
      },
      {
        "key": "io.cilium.k8s.policy.namespace",
        "value": "default",
```

Figure 6.33 – Cilium CLI shows updated policy

Use the wget command to access the nginx-cilium Service from the access Pod, as follows:

```
wget -q nginx-cilium -O -
```

The output of the following command confirms that we can access the `nginx-cilium` Service from the `access` Pod:

```
/ # wget -q --timeout=5 nginx-cilium -O -
<!DOCTYPE html>
<html>
<head>
<title>Welcome to nginx!</title>
<style>
html { color-scheme: light dark; }
body { width: 35em; margin: 0 auto;
font-family: Tahoma, Verdana, Arial, sans-serif; }
</style>
</head>
<body>
<h1>Welcome to nginx!</h1>
<p>If you see this page, the nginx web server is successfully installed and
working. Further configuration is required.</p>

<p>For online documentation and support please refer to
<a href="http://nginx.org/">nginx.org</a>.<br/>
Commercial support is available at
<a href="http://nginx.com/">nginx.com</a>.</p>

<p><em>Thank you for using nginx.</em></p>
</body>
</html>
/ #
```

Figure 6.34 – Testing access using the wget command

Now that we have completed our tasks with Cilium, we can disable Cilium CNI so that MicroK8s reverts itself to the default CNI, which is Calico CNI. The following command execution output confirms that Cilium has been disabled and MicroK8s has reverted to Calico CNI:

```
ubuntu@microk8s:~$microk8s disable cilium
Disabling Cilium
serviceaccount "cilium" deleted
serviceaccount "cilium-operator" deleted
configmap "cilium-config" deleted
clusterrole.rbac.authorization.k8s.io "cilium" deleted
clusterrole.rbac.authorization.k8s.io "cilium-operator" deleted
clusterrolebinding.rbac.authorization.k8s.io "cilium" deleted
clusterrolebinding.rbac.authorization.k8s.io "cilium-operator" deleted
daemonset.apps "cilium" deleted
deployment.apps "cilium-operator" deleted
12: cilium_vxlan: <BROADCAST,MULTICAST,UP,LOWER_UP> mtu 1440 qdisc noqu
    link/ether 0a:e4:e4:70:61:ee brd ff:ff:ff:ff:ff:ff
Deleting old cilium_vxlan link
Restarting default cni
configmap/calico-config created
```

Figure 6.35 – Disabling Cilium

The following command execution output confirms that Calico Pods are running and the default CNI is set:

```
ubuntu@microk8s:~$kubectl get pods -o wide -n kube-system
NAME                                         READY    STATUS     RESTARTS
INESS GATES
coredns-64c6478b6c-xm78p                     1/1      Running    0
e>
calico-node-nd4pn                            1/1      Running    0
e>
calico-kube-controllers-6966456d6b-94wpt     1/1      Running    0
e>
ubuntu@microk8s:~$
```

Figure 6.36 – Calico default CNI is set, and Pods are running

Cilium retains the ability to seamlessly inject security visibility and enforcement by leveraging Linux eBPF but does so in a fashion that is based on Service/Pod/container identity (rather than IP address identification, as in traditional systems) and may filter on application-layer security (for example, HTTP). As a result of decoupling security from addressing, Cilium not only makes it straightforward to apply security policies in a highly dynamic environment, but it may also provide stronger security isolation.

After looking into Cilium CNI, we can move on to Flannel CNI in the next section.

Configuring Flannel CNI

Flannel is one of the most mature open source CNI projects for Kubernetes, developed by CoreOS. Flannel is a simple network model that may be used to cover the most common Kubernetes network configuration and management scenarios. It functions by building an overlay network that assigns an internal IP address subnet to each Kubernetes cluster node. The leasing and maintenance of subnets are handled by the flanneld daemon agent, which is packaged as a single binary for easy installation and configuration on Kubernetes clusters and distributions.

Disabling the HA cluster to enable the Flannel add-on

To set Flannel as the CNI, the **high availability** (**HA**) cluster must be disabled to set the CNI as Flannel. The following command execution output confirms that the HA cluster is disabled and Flannel CNI is set:

```
ubuntu@microk8s:~$microk8s disable ha-cluster
Disabling HA will reset your cluster in a clean state.
Any running workloads will be stopped and any cluster configuration will be lost.
As this is a single node cluster and this is a destructive operation,
please use the '--force' flag.
ubuntu@microk8s:~$microk8s disable ha-cluster --force
Reverting to a non-HA setup
Generating new cluster certificates.
Waiting for node to start. .
Enabling flanneld and etcd
HA disabled
ubuntu@microk8s:~$
```

Figure 6.37 – Disabling the HA cluster to set Flannel CNI

Now that Flannel CNI is set up, we can deploy a sample application and test the network.

Flannel uses the Kubernetes etcd cluster or API to store host mappings and other network-related configurations and sustain connections between hosts/nodes via encapsulated packets after assigning IP addresses. It uses VXLAN configuration for encapsulation and communication by default, although there are a variety of backends available, including host-gw and UDP. It's also feasible to use Flannel to activate VXLAN-GBP for routing, which is required when multiple hosts are connected to the same network.

The following diagram depicts the flow of traffic between nodes using a VXLAN tunnel:

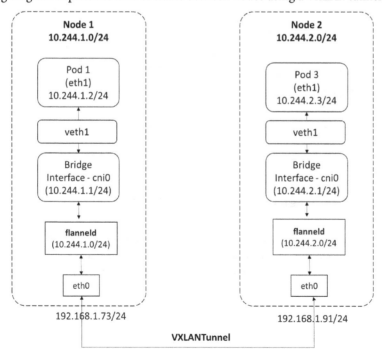

Figure 6.38 – Flannel CNI

Flannel does not include any means for encrypting encapsulated traffic by default. It does, however, enable **IP Security (IPsec)** encryption, which allows Kubernetes clusters to create encrypted tunnels between worker nodes. It is an excellent CNI plugin for novices who wish to begin their Kubernetes CNI adventure from the perspective of a cluster administrator. Until it is used to regulate traffic transfer between hosts, its simple networking model has no drawbacks.

Let's sum up what we've learned so far: we've looked at the three most popular CNI plugins: Flannel, Calico, and Cilium. When containers are built or destroyed, CNI makes it simple to configure container networking. These plugins ensure that Kubernetes' networking needs are met and cluster administrators have access to the networking functionalities they need. In the next section, we will look at some of the guidelines for choosing the right CNI provider for your requirements.

Guidelines on choosing a CNI provider

There isn't a single CNI vendor that can meet all of a project's requirements. Flannel is an excellent option for easy setup and configuration. Calico has a superior performance because it employs a BGP underlay network. Cilium uses BPF to implement an entirely different application-layer filtering model that is more focused on enterprise security.

In the following table, we compare the three most popular CNI plugins:

Parameters	Flannel	Calico	Cilium
Mode of Deployment	DaemonSet	DaemonSet	DaemonSet
Encapsulation and Routing	VXLAN	IPinIP, BGP, eBPF	VXLAN, eBPF
Support for Network Policies	No	Yes	Yes
Datastore used	Etcd	Etcd	Etcd
Encryption	Yes	Yes	Yes
Ingress Support	No	Yes	Yes

Table 6.1 – Comparison of popular CNI plugins

Relying on a single CNI provider is unnecessary because operating requirements vary widely between projects. Multiple solutions will be used and tested to meet complicated networking requirements while also giving a more dependable networking experience. We'll look at some of the most important factors to consider when selecting a CNI provider in the next section.

Key considerations when choosing a CNI provider

Calico, Flannel, and Cilium are just a few of the CNI plugins available. Let's have a look at the various aspects to consider before choosing an acceptable CNI plugin for a production environment.

You should select a plugin based on the environment in which you operate, as outlined here:

- **For virtualization environments**—There could be many network restrictions. Nodes, for example, cannot communicate directly with one another using a Layer 2 protocol; only Layer 3 features such as IP addresses are forwarded, and a host can only utilize specific IP numbers. In Overlay mode, you can only choose plugins such as Flannel-VXLAN and Calico-IPIP for a restricted underlying network.

- **For physical environments**—The underlying network is relatively unrestricted in this context. Layer 2 communication, for example, can be implemented within a switch. Plugins can be selected in **Underlay** mode in such a cluster setup. You can directly install several **network interface controllers (NICs)** onto a physical machine or virtualize hardware on NICs in Underlay mode. Routes are established in **Routing** mode using the Linux routing protocol. This prevents VXLAN encapsulation from degrading performance. You can use plugins such as Calico-BGP and Flannel-HostGW in this context.

- **For cloud environments**—The fundamental capabilities are severely limited in this context, which is a form of virtual environment. Each public cloud, on the other hand, adjusts containers for better performance and may offer APIs for configuring additional NICs or routing capabilities. For compatibility and best performance, it is recommended to use CNI plugins provided by the public cloud vendor.

After evaluating the environmental constraints, you may have a better sense of which plugins can and cannot be used. In the next section, we will look at business requirements.

Based on business requirements, functional criteria could also dictate your plugin options. Here are some factors that should be considered:

- **Security requirements**—Kubernetes includes NetworkPolicy, which lets you set up rules to support policies such as whether to allow access between Pods. NetworkPolicy declaration is not supported by all CNI plugins. Calico is a good option if you need NetworkPolicy support.

- **Connection to resources within and outside the cluster**—Applications running on VMs or physical machines can't all be moved to a containerized environment at the same time. As a result, IP address connectivity between VMs or physical machines and containers must be configured by interconnecting or deploying them at the same layer. In these kinds of scenarios, choose a plugin in Underlay mode. The Calico-BGP plugin, for example, allows Pods and legacy VMs or physical machines to share a layer. Even though containers are in a different CIDR block than historical VMs or physical machines, Calico-BGP can be used to publish BGP routes to original routers, allowing VMs and containers to communicate.

- **Service discovery and load-balancing capabilities**—Kubernetes provides services such as service discovery and load balancing. These two features are not available in all CNI plugins. In Underlay mode, the NIC of a Pod is either the Underlay hardware or is virtualized and introduced into a container via hardware for various plugins. As a result, NIC traffic cannot be forwarded to the namespace in which the host resides, and you won't be able to use the rules that `kube-proxy` sets up on the host. In this instance, the plugin is unable to use Kubernetes' service discovery capabilities. Choose a plugin in **Underlay** mode that enables service discovery and load balancing if you need these features.

Now that we've gone over the business criteria for selecting a plugin, we can move on to the performance requirements.

Based on performance requirements, Pod creation speed and Pod network performance could be used to gauge performance. Depending on the implementation modes, there may be a performance loss. Here are some of the considerations when choosing a plugin:

- **Pod creation speed**—For example, there could be scenarios to build and configure more network resources when you need to scale out immediately during a business peak scenario. In the case of the CNI plugin with Overlay mode, you can easily scale up Pods because the plugin implements virtualization on nodes, and creating Pods is as simple as calling kernel interfaces. In the case of Underlay mode, it must first generate underlying network resources, which slows down the Pod-generation process. Hence, when you need to quickly scale out Pods or build a large number of Pods, choose an **Overlay** mode plugin.

- **Pod network performance**—Metrics such as inter-Pod network forwarding, network bandwidth, and **pulse-per-second** (**PPS**) latency are used to assess Pod network performance. Plugins in **Overlay** mode will give lesser performance than plugins in Underlay modes since the former implement virtualization on nodes and encapsulate packets. As a result, if you need excellent network performance in scenarios such as **machine learning** (**ML**) and big-data scenarios, don't select a plugin in Overlay mode; instead, use a CNI plugin in Underlay mode.

Summary

In this chapter, we looked at how networking is handled in a Kubernetes cluster. We also learned how CNI supports dynamic networking resource setup, such as network configuration, IP address provisioning, and multi-host communication. We learned how CNI automatically configures networks between Pods using either an underlay or an overlay network.

We've also covered how to use Calico, Cilium, and Flannel CNI plugins to network the cluster. We discovered the advantages and disadvantages of each CNI. We also discovered that no single CNI vendor was capable of meeting all of a project's requirements. Flannel is an excellent solution for easy setup and configuration. Calico has a superior performance because it employs a BGP underlay network. BPF is used by Cilium to create an application-layer filtering approach that is more focused on enterprise security. We've gone through some of the most important factors to consider when selecting a CNI Service.

In the next chapter, we'll continue with our next use case, which is about exposing your Services outside the cluster.

## 7

Setting Up MetalLB and Ingress for Load Balancing

In the last chapter, we have looked at how Kubernetes network model works and learned how to use Calico, Cilium, and Flannel CNI plugins to network the cluster. We've also gone through some of the most important factors to consider when choosing a CNI provider.

We should revisit the Kubernetes Service abstraction mechanism from the first chapter before diving into **MetalLB** load-balancer and Ingress concepts for load balancing. Kubernetes **Services**, in simple terms, connect a group of Pods to an abstracted Service name and IP address. Discovery and routing between Pods are provided by the Services. Services, for example, connect an application's frontend to its backend, which are both deployed in different cluster deployments.

The most common types of Services are listed here:

- **ClusterIP**: This is the default type, which exposes the Service via the cluster's internal IP address. These Services are only accessible within the cluster.

- **NodePort**: A static port on each node's IP address is used to expose a Service. To route traffic to the `NordPort` service, a `ClusterIP` Service is automatically created.

- **LoadBalancer**: The `LoadBalancer` type of service will create a load balancer and expose the Service externally. It will also automatically create `ClusterIP` and `NodePort` Services and route traffic accordingly.

- **ExternalName**: Maps a Service to a predefined `externalName ex.sampleapp.test.com` field by returning a value for the **Canonical Name (CNAME)** record.

The most common types of Services are depicted in the following diagram:

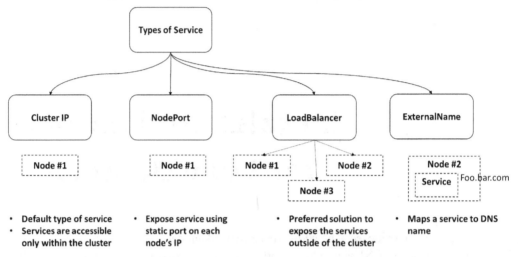

Figure 7.1 – Common types of Services

Let's get into the intricacies of MetalLB and Ingress configuration now that we've covered the basics. In this chapter, we're going to cover the following main topics:

- Overview of MetalLB and Ingress
- Configuring MetalLB to load balance across the cluster
- Configuring Ingress to expose Services outside the cluster
- Guidelines on choosing the right load balancer for your applications

Overview of MetalLB and Ingress

Despite its widespread adoption, Kubernetes does not offer a load balancer implementation. If your Kubernetes cluster is running on a cloud platform such as Azure, **Amazon Web Services** (**AWS**), or **Google Cloud Platform** (**GCP**), the cluster can use the underlying cloud platform's load balancer implementation through the cloud-controller-manager API. However, not all Kubernetes clusters are hosted in the cloud. Kubernetes may be installed on bare-metal machines as well, which is most common in the edge computing world. When load balancers are created in this situation, they will remain in a `Pending` status indefinitely.

The `NodePort` and `externalIPs` Services are the only options for bringing user traffic into bare-metal clusters. Both strategies have considerable drawbacks when it comes to output. MetalLB solves this problem by providing an implementation of a network load balancer that connects with conventional network equipment, allowing external Services on bare-metal clusters.

In a nutshell, MetalLB enables you to establish `LoadBalancer` Kubernetes Services in clusters that aren't hosted on a cloud provider. Address allocation and external announcement are two characteristics that work together to deliver this Service. We will now see these in more detail, as follows:

- **Address allocation**: MetalLB cannot generate IP addresses on its own; instead, we must provide it with IP address pools from which it can draw. As Services come and go, MetalLB will take care of assigning and unassigning individual addresses, but it will only ever distribute IPs that are part of its preset pools. Setting up IP address pools is based on the environment we have; for example, if you're running a bare-metal cluster in a colocation facility, your hosting provider may provide IP addresses for lease, or if you're running on a private **local area network** (**LAN**), you could choose a range of IPs from one of the private addresses spaces.

- **External announcement**: After assigning an external IP address to a Service, MetalLB must notify the network outside the cluster that the IP address is residing in the cluster. MetalLB accomplishes this using conventional routing protocols such as **Address Resolution Protocol** (**ARP**), **Neighbor Discovery Protocol** (**NDP**), or **Border Gateway Protocol** (**BGP**). In a **Layer 2** (**L2**) mode such as ARP/NDP, one node in the cluster takes ownership of the Service and makes those IPs visible on the local network using standard address discovery protocols (ARP for IPv4; NDP for IPv6); whereas in the BGP mode, all nodes in the cluster create BGP peering sessions with adjacent routers that you control in BGP mode and inform those routers how to forward traffic to the Service IPs. BGP's policy mechanisms enable genuine load balancing across several nodes as well as fine-grained traffic control.

Another option is to utilize **Ingress** (Kubernetes object) to expose your Service. Although it acts as the cluster's entrance point, Ingress is not a Service type.

The workings of Ingress are depicted in the following diagram:

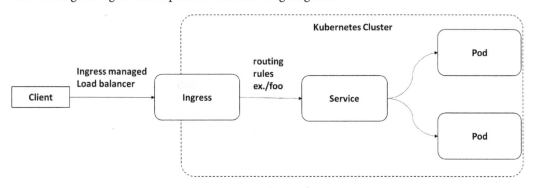

Figure 7.2 – Workings of Ingress

The Ingress controller aids in the consolidation of various applications' routing rules into a single entity. With the help of `NodePort` or `LoadBalancer`, the Ingress controller is exposed to the outside world. It's more suited for internal load balancing of **HyperText Transfer Protocol** (**HTTP**) or **HTTP Secure** (**HTTPS**) traffic to your deployed Services utilizing load balancers such as `nginx` or

HAProxy. For **Transmission Control Protocol** (**TCP**) or **User Datagram Protocol** (**UDP**) traffic, we may still utilize a `LoadBalancer` kind of service, and MetalLB comes to our rescue in such situations.

In the next section, we'll go over how to set up MetalLB as a load balancer for your cluster.

Configuring MetalLB to load balance across the cluster

Now that we are clear on MetalLB concepts, we will dive into the steps of configuring MetalLB to load balance across the cluster. The following diagram depicts our Raspberry Pi cluster setup:

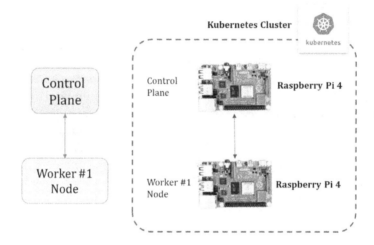

Figure 7.3 – MicroK8s Raspberry Pi cluster

Now that we know what we want to do, let's look at the requirements.

Requirements

Before you begin, here are the prerequisites that are needed for building a Raspberry Pi Kubernetes cluster and for the configuration of a MetalLB load balancer:

- A microSD card (4 **gigabytes** (**GB**) minimum; 8 GB recommended)
- A computer with a microSD card drive
- A Raspberry Pi 2, 3, or 4 (one or more)
- A micro-USB power cable (USB-C for the Pi 4)
- A Wi-Fi network or an Ethernet cable with an internet connection
- (Optional) A monitor with a **High-Definition Multimedia Interface** (**HDMI**) interface

- (Optional) An HDMI cable for the Pi 2 and 3 and a micro-HDMI cable for the Pi 4
- (Optional) A **Universal Serial Bus (USB)** keyboard

Now that we've established what the requirements are, we'll go on to the step-by-step instructions on how to complete the process.

Step 1 – Creating a MicroK8s Raspberry Pi cluster

Please follow the steps that we covered in *Chapter 5, Creating and Implementing Updates on Multi-Node Raspberry Pi Kubernetes Clusters*, to create a MicroK8s Raspberry Pi cluster. Here's a quick refresher:

1. Installing the **operating system (OS)** image to a **Secure Digital (SD)** card:

    A. Configuring Wi-Fi access settings

    B. Configuring remote access settings

    C. Configuring control group settings

    D. Configuring hostname

2. Installing and configuring MicroK8s

3. Adding a worker node

A fully functional multi-node Kubernetes cluster should look like the one shown in the following screenshot. To summarize, we have installed MicroK8s on the Raspberry Pi boards and joined multiple deployments to form a cluster. We have also added nodes to the cluster:

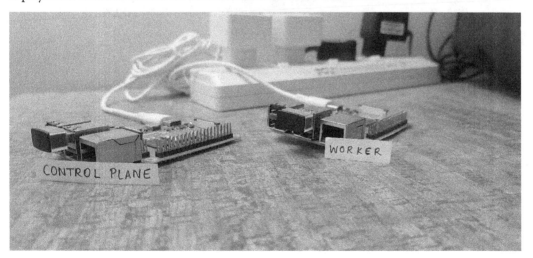

Figure 7.4 – Fully functional MicroK8s Raspberry Pi cluster

We can now go to the next step of enabling the MetalLB add-on, as we have a fully functional cluster.

Step 2 – Enabling the MetalLB add-on

MicroK8s supports a variety of add ons (`https://microk8s.io/docs/addons`) which are pre-packaged components that provide additional capabilities for your Kubernetes cluster.

These are easy to set up with the following command:

```
microk8s enable <<add-on>>
```

Use the following command to enable the MetalLB load balancer:

```
microk8s enable metallb <<list of IP address>>
```

The following command execution output indicates the MetalLB add-on has been enabled successfully:

```
ubuntu@controlplane:~$ microk8s enable metallb 192.168.1.10-192.168.1.15
Enabling MetalLB
Applying Metallb manifest
namespace/metallb-system created
secret/memberlist created
Warning: policy/v1beta1 PodSecurityPolicy is deprecated in v1.21+, unavailable in v1.25+
podsecuritypolicy.policy/controller created
podsecuritypolicy.policy/speaker created
serviceaccount/controller created
serviceaccount/speaker created
clusterrole.rbac.authorization.k8s.io/metallb-system:controller created
clusterrole.rbac.authorization.k8s.io/metallb-system:speaker created
role.rbac.authorization.k8s.io/config-watcher created
role.rbac.authorization.k8s.io/pod-lister created
clusterrolebinding.rbac.authorization.k8s.io/metallb-system:controller created
clusterrolebinding.rbac.authorization.k8s.io/metallb-system:speaker created
rolebinding.rbac.authorization.k8s.io/config-watcher created
rolebinding.rbac.authorization.k8s.io/pod-lister created
Warning: spec.template.spec.nodeSelector[beta.kubernetes.io/os]: deprecated since v1.14;
daemonset.apps/speaker created
deployment.apps/controller created
configmap/config created
MetalLB is enabled
ubuntu@controlplane:~$
```

Figure 7.5 – Enabling MetalLB add-on

We've instructed MetalLB to give out addresses in the 192.168.1.10 - 192.168.1.15 range with this command. To check a list of available and installed add-ons, use the status command, as follows:

```
microk8s status
```

> **Note**
>
> Alternatively, you can use **Classless Inter-Domain Routing** (**CIDR**) notation. For example, in my network, I use the `192.168.2.1/24` subnet, and I opted to give MetalLB half of the IPs. IP numbers `192.168.2.1` to `192.168.2.126` make up the first part of the subnet. A /25 subnet can be used to represent this range: `192.168.2.1/25`.
>
> A /25 subnet can also be used to represent the second half of the network—for example, `192.168.2.128/25`. Each half has 126 IP addresses.
>
> Make sure you choose subnets that are appropriate for your network and that your router and MetalLB are configured correctly.

The following command execution output indicates (refer to the highlighted portions) that the MetalLB add-on has been enabled:

```
ubuntu@controlplane:~$ microk8s status
microk8s is running
high-availability: no
  datastore master nodes: 127.0.0.1:19001
  datastore standby nodes: none
addons:
  enabled:
    dns                      # CoreDNS
    ha-cluster               # Configure high availability on the current node
    metallb                  # Loadbalancer for your Kubernetes cluster
  disabled:
    dashboard                # The Kubernetes dashboard
    dashboard-ingress        # Ingress definition for Kubernetes dashboard
```

Figure 7.6 – MicroK8s status

For its components, MetalLB uses the `metallb-system` namespace. To verify all components are running, use the following command:

```
kubectl get all -n metallb-system
```

The following command execution output indicates that all components are in a `Running` state:

```
ubuntu@controlplane:~$ kubectl get all -n metallb-system
NAME                             READY   STATUS    RESTARTS   AGE
pod/speaker-zwz98                1/1     Running   0          3m9s
pod/controller-558b7b958-t4r78   1/1     Running   0          3m9s

NAME                      DESIRED   CURRENT   READY   UP-TO-DATE   AVAILABLE   NODE SELECTOR              AGE
daemonset.apps/speaker    1         1         1       1            1           beta.kubernetes.io/os=linux   3m9s

NAME                          READY   UP-TO-DATE   AVAILABLE   AGE
deployment.apps/controller    1/1     1            1           3m9s

NAME                                    DESIRED   CURRENT   READY   AGE
replicaset.apps/controller-558b7b958    1         1         1       3m9s
ubuntu@controlplane:~$
```

Figure 7.7 – Components of MetalLB and their status

The components that you can see in the preceding command execution output of MetalLB are outlined in more detail here:

- `metallb-system/controller` (deployment): This is the IP address assignment controller for the entire cluster.
- `metallb-system/speaker` (DaemonSet): This component communicates with Services using the protocol(s) of your choice.

Now that we've activated the MetalLB add-on, the next step is to launch an example application and see whether it can load balance.

Add-ons that have been enabled can be disabled at any time by utilizing the `disable` command, as follows:

```
microk8s disable <<add-on>>
```

At this point, you have a fully functional multi-node Kubernetes cluster with the MetalLB add-on enabled.

> **Note**
> Between nodes, port 7946 (TCP and UDP) must be permitted. Additionally, ensure that no other software is running on port 7946 on the nodes before installing MetalLB.

Step 3 – Deploying a sample containerized application

In this step, we will be deploying the following Apache web server deployment on our multi-node MicroJ8s cluster setup, as follows:

```
apiVersion: v1
kind: Namespace
metadata:
  name: web
---
apiVersion: apps/v1
kind: Deployment
metadata:
  name: web-server
  namespace: web
spec:
```

```
  selector:
    matchLabels:
      app: web
  template:
    metadata:
      labels:
        app: web
    spec:
      containers:
      - name: httpd
        image: httpd:2.4-alpine
        ports:
        - containerPort: 80
---
apiVersion: v1
kind: Service
metadata:
  name: web-server-service
  namespace: web
spec:
  selector:
    app: web
  ports:
    - protocol: TCP
      port: 80
      targetPort: 80
  type: LoadBalancer
```

Before we deploy the web server, let's verify the cluster nodes are ready by using the following command:

```
kubectl get nodes
```

The following command execution output shows that all the nodes are in a `Ready` state. We are now ready to start:

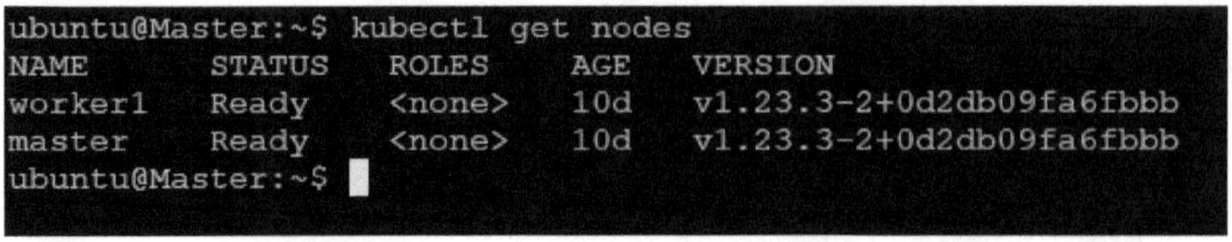

```
ubuntu@Master:~$ kubectl get nodes
NAME       STATUS    ROLES     AGE     VERSION
worker1    Ready     <none>    10d     v1.23.3-2+0d2db09fa6fbbb
master     Ready     <none>    10d     v1.23.3-2+0d2db09fa6fbbb
ubuntu@Master:~$
```

Figure 7.8 – Checking whether nodes are in a Ready state

The following command will deploy the web server application:

```
kubectl apply -f webserver-deploy.yaml
```

The following command execution output indicates that there is no error in the deployment, and in the next steps, we can verify the same using the `get deployments` command:

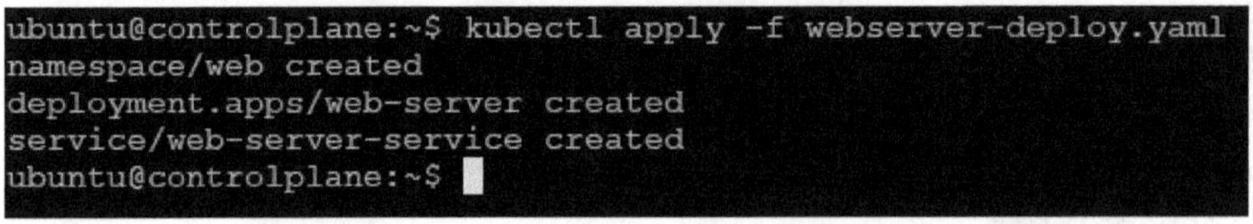

```
ubuntu@controlplane:~$ kubectl apply -f webserver-deploy.yaml
namespace/web created
deployment.apps/web-server created
service/web-server-service created
ubuntu@controlplane:~$
```

Figure 7.9 – Deploying the web server

Check the status of the deployment to verify the application has been deployed and is running by using the following command:

```
kubectl get deployments -n web
```

The following command execution output confirms that the deployment is `Ready`:

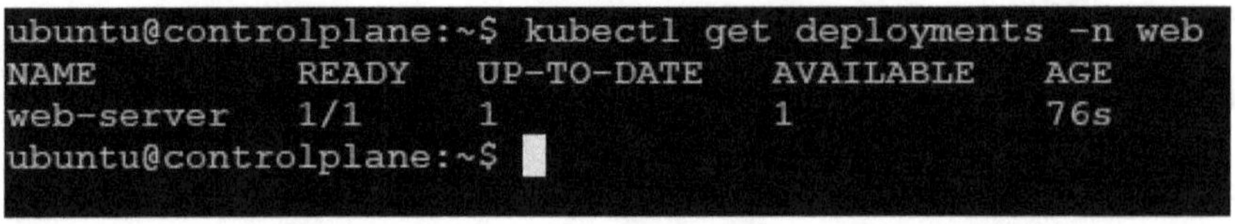

```
ubuntu@controlplane:~$ kubectl get deployments -n web
NAME          READY    UP-TO-DATE    AVAILABLE    AGE
web-server    1/1      1             1            76s
ubuntu@controlplane:~$
```

Figure 7.10 – Confirming that the deployment is ready

Now that we have the sample web server application ready, we can test the load-balancing mechanism in the next part.

Step 4 – Verifying the load balancer mechanism

To summarize, we've deployed a sample web server application. We'll now test the load-balancer mechanism in this section.

Use the following command to see whether your load balancer has been allocated an external IP and port:

```
kubectl get all -n web
```

The following command execution output (refer to the highlighted portions) shows that an external IP and port have been allocated:

```
ubuntu@controlplane:~$ kubectl get all -n web
NAME                                READY   STATUS      RESTARTS   AGE
pod/web-server-54c7c6444c-vn2fj     1/1     Running     0          101s

NAME                          TYPE           CLUSTER-IP       EXTERNAL-IP     PORT(S)        AGE
service/web-server-service    LoadBalancer   10.152.183.214   192.168.1.10    80:31323/TCP   100s

NAME                          READY   UP-TO-DATE   AVAILABLE   AGE
deployment.apps/web-server    1/1     1            1           101s

NAME                                     DESIRED   CURRENT   READY   AGE
replicaset.apps/web-server-54c7c6444c    1         1         1       101s
ubuntu@controlplane:~$
```

Figure 7.11 – Checking whether external IP and port have been allocated

Now that an external IP has been allocated, we can access the application using the external IP address from any of the nodes or from the external network, as follows:

```
curl 192.168.1.10
```

If you have followed the preceding steps, you should be able to see the `<html><body><h1>It works!</h1></body></html>` Apache web server output, as in the following command execution output:

```
ubuntu@controlplane:~$ curl 192.168.1.10
<html><body><h1>It works!</h1></body></html>
ubuntu@controlplane:~$
```

Figure 7.12 – Apache web server output

Let's scale the deployment to see whether our load balancer is still working properly. To do so, run the `kubectl scale` command, as follows:

```
kubectl scale deployments/web-server --replicas=5 -n web
```

The following command execution output confirms that there is no error in the deployment, and in the next steps, we can verify that the deployment has five Pods:

```
ubuntu@controlplane:~$ kubectl scale deployments/web-server --replicas=5 -n web
deployment.apps/web-server scaled
ubuntu@controlplane:~$ 
```

Figure 7.13 – Scaling the deployment

Use the following command to check the status of the deployment:

```
kubectl get deployments -n web
```

The following command output shows that the command was successfully run, and the deployment has been updated:

```
ubuntu@controlplane:~$ kubectl get deployments -n web
NAME          READY     UP-TO-DATE    AVAILABLE    AGE
web-server    5/5       5             5            3m34s
ubuntu@controlplane:~$ 
```

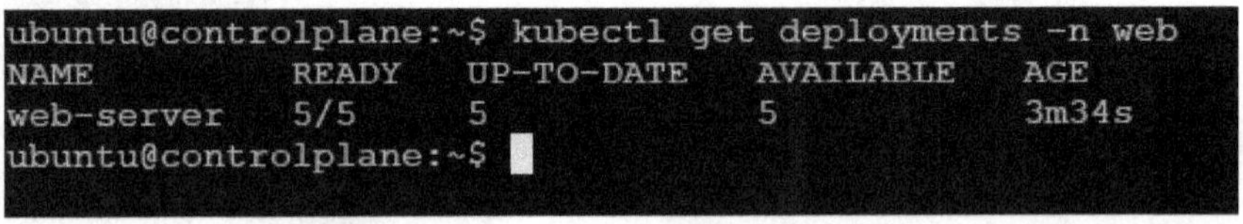

Figure 7.14 – Checking deployment state

Now that the deployment has been updated, let's check how the Pods are distributed across the nodes, using the following command:

```
kubectl get pods -n web -o wide
```

The following command execution output indicates that three Pods are running on the master node, and two of them are running on worker1:

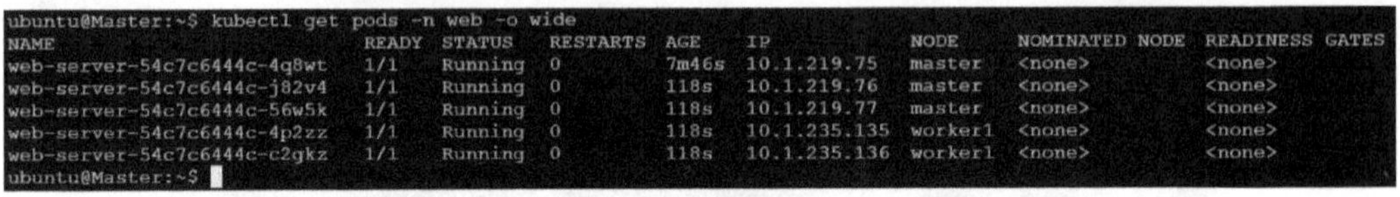

```
ubuntu@Master:~$ kubectl get pods -n web -o wide
NAME                          READY  STATUS   RESTARTS  AGE    IP             NODE      NOMINATED NODE  READINESS GATES
web-server-54c7c6444c-4q8wt   1/1    Running  0         7m46s  10.1.219.75    master    <none>          <none>
web-server-54c7c6444c-j82v4   1/1    Running  0         118s   10.1.219.76    master    <none>          <none>
web-server-54c7c6444c-56w5k   1/1    Running  0         118s   10.1.219.77    master    <none>          <none>
web-server-54c7c6444c-4p2zz   1/1    Running  0         118s   10.1.235.135   worker1   <none>          <none>
web-server-54c7c6444c-c2gkz   1/1    Running  0         118s   10.1.235.136   worker1   <none>          <none>
ubuntu@Master:~$ 
```

Figure 7.15 – How Pods are distributed across the nodes

Let's check whether there is any change in the external IP and ports using the following command:

```
kubectl get all -n web
```

The following command execution output shows that there is no change in the allocated external IP and port:

```
ubuntu@controlplane:~$ kubectl get all -n web
NAME                             READY   STATUS    RESTARTS   AGE
pod/web-server-54c7c6444c-vn2fj   1/1    Running   0          4m37s
pod/web-server-54c7c6444c-8hk4v   1/1    Running   0          93s
pod/web-server-54c7c6444c-mmzjp   1/1    Running   0          93s
pod/web-server-54c7c6444c-9vt7q   1/1    Running   0          93s
pod/web-server-54c7c6444c-rpv25   1/1    Running   0          92s

NAME                          TYPE           CLUSTER-IP       EXTERNAL-IP    PORT(S)        AGE
service/web-server-service    LoadBalancer   10.152.183.214   192.168.1.10   80:31323/TCP   4m36s

NAME                          READY   UP-TO-DATE   AVAILABLE   AGE
deployment.apps/web-server    5/5     5            5           4m37s

NAME                                    DESIRED   CURRENT   READY   AGE
replicaset.apps/web-server-54c7c6444c   5         5         5       4m37s
ubuntu@controlplane:~$
```

Figure 7.16 – Rechecking for any change in external IP and port

Let's use the same `curl` command to access the application and verify it's working, as follows:

```
ubuntu@controlplane:~$ curl 192.168.1.10
<html><body><h1>It works!</h1></body></html>
ubuntu@controlplane:~$
```

Figure 7.17 – Apache web server output

The command execution output confirms that the load balancer is working properly. Even though the Pods are spread throughout the nodes, our web server application is able to effectively serve user requests.

The following diagram depicts the MetalLB load-balancing functionality. MetalLB implements the LoadBalancer Kubernetes Service. When a LoadBalancer Service is requested externally, MetalLB assigns an IP address from the preset range to the client and informs the network that the IP is residing in the cluster:

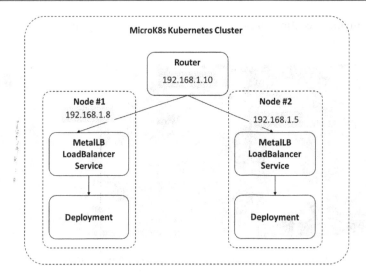

Figure 7.18 – MetalLB load-balancing functionality

MetalLB is configured in L2 mode by default in MicroK8s. The following command can be used to confirm this:

```
kubectl describe configmap config -n metallb-system
```

The following command execution output confirms that MetalLB is in L2 mode, and we can see the IP range as well:

```
ubuntu@controlplane:~$ kubectl describe configmap config -n metallb-system
Name:          config
Namespace:     metallb-system
Labels:        <none>
Annotations:   <none>

Data
====
config:
----
address-pools:
- name: default
  protocol: layer2
  addresses:
  - 192.168.1.10-192.168.1.15


BinaryData
====

Events:   <none>
ubuntu@controlplane:~$
```

Figure 7.19 – MetalLB ConfigMap

To set the MetalLB in BGP mode, `ConfigMap config` needs to be reconfigured to set the operation mode as `BGP` and the external IP address range.

The speakers in BGP mode create a BGP peering with routers outside of the cluster and instruct those routers on how to redirect traffic to the Service IPs. BGP's policy mechanisms enable genuine load balancing across several nodes, as well as fine-grained traffic control.

You can utilize ordinary router hardware with BGP as a load-balancing method. It does, however, have some drawbacks. More information regarding these limits, as well as ways to overcome them, may be found on the MetalLB BGP documentation page (`https://metallb.universe.tf/configuration/#bgp-configuration`).

To summarize, MetalLB enables you to establish Kubernetes `LoadBalancer` Services without requiring your cluster to be deployed on a cloud platform. MetalLB provides two modes of operation: a basic L2 mode that requires no external hardware or configuration, and a BGP mode that is more robust and production-ready but necessitates more network setup tasks. In the next section, we will look at how to use the `Ingress` method for load-balancing configuration.

Configuring Ingress to expose Services outside the cluster

As we discussed in the *Overview of MetalLB and Ingress* section, Ingress offers HTTP and HTTPS routes to Services within the cluster from outside the cluster. Rules defined on Ingress control traffic routing. NGINX Ingress Controller is a common Kubernetes Ingress and the default Ingress controller for MicroK8s as well.

Another option is employing a load balancer such as MetalLB that can be deployed in the same Kubernetes cluster, and the Services can then be exposed to an external network.

A diagrammatic illustration of both approaches is shown here:

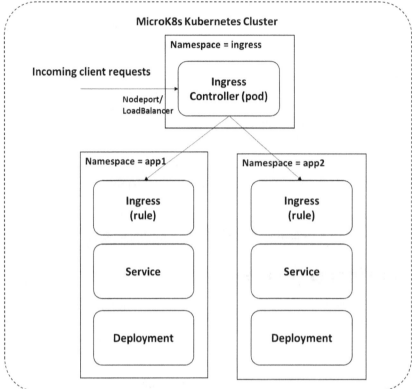

Figure 7.20 – Ingress load-balancing functionality

Both options are discussed in more depth in the following sections.

Option 1 – Using the Ingress NodePort method

For this option and the next, we'll use the same MicroK8s Raspberry Pi cluster that we built for the MetalLB setup.

Step 1 – Enabling the Ingress add-on

The Ingress add-on can be enabled using the same command that we used to enable MetalLB, as illustrated here:

```
microk8s enable ingress
```

The following command execution output indicates the Ingress add-on has been enabled successfully:

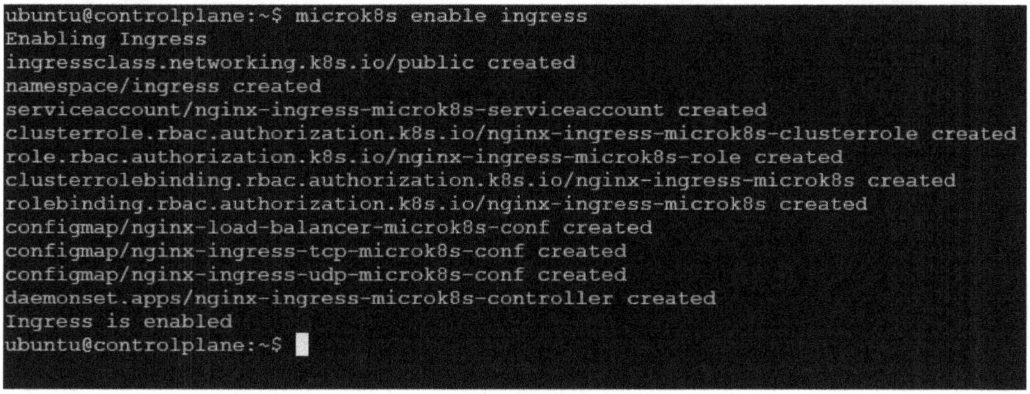

```
ubuntu@controlplane:~$ microk8s enable ingress
Enabling Ingress
ingressclass.networking.k8s.io/public created
namespace/ingress created
serviceaccount/nginx-ingress-microk8s-serviceaccount created
clusterrole.rbac.authorization.k8s.io/nginx-ingress-microk8s-clusterrole created
role.rbac.authorization.k8s.io/nginx-ingress-microk8s-role created
clusterrolebinding.rbac.authorization.k8s.io/nginx-ingress-microk8s created
rolebinding.rbac.authorization.k8s.io/nginx-ingress-microk8s created
configmap/nginx-load-balancer-microk8s-conf created
configmap/nginx-ingress-tcp-microk8s-conf created
configmap/nginx-ingress-udp-microk8s-conf created
daemonset.apps/nginx-ingress-microk8s-controller created
Ingress is enabled
ubuntu@controlplane:~$
```

Figure 7.21 – Enabling Ingress add-on

Now that the Ingress add-on has been enabled, the next step is to deploy a sample application to test the load-balancing functionality.

Step 2 – Deploying a sample containerized application

We'll apply the following `whoami` deployment on our multi-node MicroK8s cluster, which is a Tiny Go web server that prints OS information and HTTP requests to output:

```
apiVersion: apps/v1
kind: Deployment
metadata:
  name: whoami-deployment
spec:
  replicas: 1
  selector:
    matchLabels:
      app: whoami
  template:
    metadata:
      labels:
        app: whoami
    spec:
      containers:
      - name: whoami-container
```

```
            image: containous/whoami
---
apiVersion: v1
kind: Service
metadata:
  name: whoami-service
spec:
  ports:
  - name: http
    targetPort: 80
    port: 80
  selector:
    app: whoami
---
apiVersion: networking.k8s.io/v1
kind: Ingress
metadata:
  name: whoami-ingress
spec:
  rules:
  - http:
      paths:
      - path: /whoami
        pathType: Exact
        backend:
          service:
            name: whoami-service
            port:
              number: 80
```

The following command will deploy the whoami application:

```
kubectl apply -f whoami-deployment.yaml
```

The following command execution output indicates that there is no error in the deployment, and in the next steps, we can verify this by using the get deployments command:

```
ubuntu@controlplane:~$ kubectl apply -f whoami-deployment.yaml
deployment.apps/whoami-deployment created
service/whoami-service created
ingress.networking.k8s.io/whoami-ingress created
ubuntu@controlplane:~$
```

Figure 7.22 – Deploying the whoami application

Check the status of the deployment to verify the application has been deployed and is running by using the following command:

```
kubectl get deployments
```

The following command execution output indicates that the deployment is Ready:

```
ubuntu@controlplane:~$ kubectl get deployments
NAME                READY   UP-TO-DATE   AVAILABLE   AGE
whoami-deployment   1/1     1            1           69s
ubuntu@controlplane:~$
```

Figure 7.23 – Checking the deployment status

Use the following command to check whether an Ingress object has been created:

```
kubectl get ingress
```

The output of the preceding command indicates that an Ingress object with the name whoami-ingress has been created:

```
ubuntu@controlplane:~$ kubectl get ingress
NAME             CLASS    HOSTS   ADDRESS     PORTS   AGE
whoami-ingress   public   *       127.0.0.1   80      91s
ubuntu@controlplane:~$
```

Figure 7.24 – Checking the Ingress object that we created

Use the `describe` command to view detailed information about the Ingress object we just created:

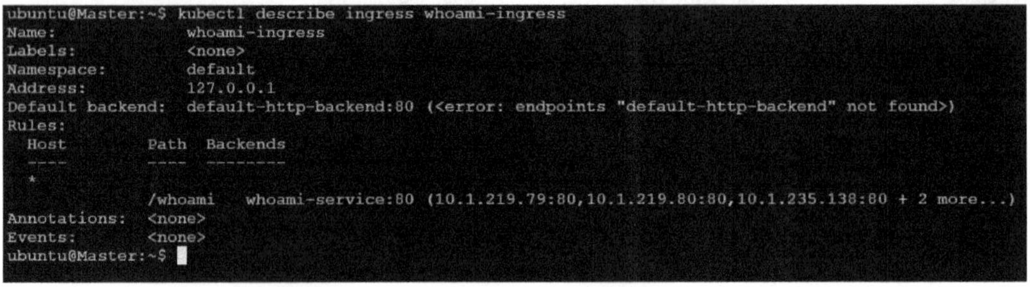

```
ubuntu@Master:~$ kubectl describe ingress whoami-ingress
Name:            whoami-ingress
Labels:          <none>
Namespace:       default
Address:         127.0.0.1
Default backend: default-http-backend:80 (<error: endpoints "default-http-backend" not found>)
Rules:
  Host        Path  Backends
  ----        ----  --------
  *
              /whoami    whoami-service:80 (10.1.219.79:80,10.1.219.80:80,10.1.235.138:80 + 2 more...)
Annotations:  <none>
Events:       <none>
ubuntu@Master:~$
```

Figure 7.25 – Ingress describe command to check details on Ingress object created

From the command execution output, here is what each field represents:

- `Host`: Because no host is mentioned in the previous output, the rule applies to all inbound HTTP traffic via the provided IP address. The rules apply to that site if a host (for example, `foo.com`) is supplied.

- `Path`: List of routes (for example, `/whoami`) with a backend described by a `service.name` and a `service.port.name` or `service.port.number`. Before the load balancer distributes traffic to the specified Service, the host and path must match the content of an incoming request.

- `Backends`: The Service describes a backend as a combination of Service and port names. The mentioned backend receives HTTP (and HTTPS) requests to the Ingress object that matches the rule's host and path.

For more information on Ingress, please refer to the Ingress Kubernetes documentation at the following link: `https://kubernetes.io/docs/concepts/services-networking/ingress/`.

Step 3 – Verifying the load balancer mechanism

Use the `curl` command to check whether you can access the application, as illustrated here:

```
ubuntu@worker1:~$ curl 192.168.1.8/whoami
Hostname: whoami-deployment-57fb67548c-mh4zp
IP: 127.0.0.1
IP: ::1
IP: 10.1.235.138
IP: fe80::1c3a:79ff:fe9d:bf29
RemoteAddr: 10.1.219.78:40750
GET /whoami HTTP/1.1
Host: 192.168.1.8
User-Agent: curl/7.68.0
Accept: */*
X-Forwarded-For: 192.168.1.7
X-Forwarded-Host: 192.168.1.8
X-Forwarded-Port: 80
X-Forwarded-Proto: http
X-Forwarded-Scheme: http
X-Real-Ip: 192.168.1.7
X-Request-Id: 03a64ad7607dee3bd7e28604561d62e9
X-Scheme: http
```

Figure 7.26 – Accessing the deployed application

Let's scale the deployment to see whether our load balancer is working properly. To do so, run the kubectl scale command, as follows:

```
kubectl scale deployments/whoami-deployment --replicas=5
```

The following command execution output confirms that there is no error in the deployment, and in the next steps, let's check how the Pods are distributed across the nodes:

```
ubuntu@controlplane:~$ kubectl scale deployments/whoami-deployment --replicas=5
deployment.apps/whoami-deployment scaled
ubuntu@controlplane:~$
```

Figure 7.27 – Scaling the deployment

The following command execution output indicates that two Pods are running on the master node, and three of them are running on worker1:

```
ubuntu@Master:~$ kubectl get pods -o wide
NAME                                 READY   STATUS    RESTARTS   AGE     IP             NODE      NOMINATED NODE   READINESS GATES
whoami-deployment-57fb67548c-mh4zp   1/1     Running   0          6m52s   10.1.235.138   worker1   <none>           <none>
whoami-deployment-57fb67548c-dnsc5   1/1     Running   0          2m43s   10.1.235.139   worker1   <none>           <none>
whoami-deployment-57fb67548c-gpp6h   1/1     Running   0          2m43s   10.1.235.140   worker1   <none>           <none>
whoami-deployment-57fb67548c-lsk6q   1/1     Running   0          2m43s   10.1.219.79    master    <none>           <none>
whoami-deployment-57fb67548c-zndn7   1/1     Running   0          2m43s   10.1.219.80    master    <none>           <none>
ubuntu@Master:~$
```

Figure 7.28 – Checking how the Pods are distributed across nodes

Let's use the same `curl` command to access the application and verify it's working, as follows:

```
ubuntu@worker1:~$ curl 192.168.1.8/whoami
Hostname: whoami-deployment-57fb67548c-1sk6q
IP: 127.0.0.1
IP: ::1
IP: 10.1.219.79
IP: fe80::1481:7ff:fe9b:1578
RemoteAddr: 10.1.219.78:51910
GET /whoami HTTP/1.1
Host: 192.168.1.8
User-Agent: curl/7.68.0
Accept: */*
X-Forwarded-For: 192.168.1.7
X-Forwarded-Host: 192.168.1.8
X-Forwarded-Port: 80
X-Forwarded-Proto: http
X-Forwarded-Scheme: http
X-Real-Ip: 192.168.1.7
X-Request-Id: 7aee251014f64f663bf54c5af19b5b43
X-Scheme: http
```

Figure 7.29 – Rechecking whether the load balancer is working properly

The command execution output confirms that the load balancer is working properly. Even though the Pods are spread throughout the nodes, our whoami application is able to effectively serve user requests.

Option 2 – Using Ingress and a load balancer

In this method, we must have a load balancer for the Kubernetes cluster in order to proceed. Since we already have an installed MetalLB load balancer configured, we will be reusing it. However, we will have to define a simple load balancer Service for our sample whoami application to make sure it's acquiring the external IP and port.

Here is the code for the simple load balancer Service deployment of the whoami application:

```
apiVersion: v1
kind: Service
metadata:
  name: metallb-load-balancer
spec:
  selector:
```

```
        app: whoami
    ports:
      - protocol: TCP
        port: 80
        targetPort: 80
    type: LoadBalancer
```

The following command will deploy the preceding application:

```
kubectl apply -f loadbalancer.yaml
```

The following command output shows that the command was successfully run and a load balancer has been created. We will confirm the same in the next steps:

```
ubuntu@controlplane:~$ kubectl apply -f loadbalancer.yaml
service/metallb-load-balancer created
ubuntu@controlplane:~$
```

Figure 7.30 – Created load-balancer Service

Use the following command to check whether a load balancer Service has been created and an external IP and port have been allocated. You won't acquire an IP address for EXTERNAL-IP if a load balancer isn't available. Instead, it's marked as <pending>. In this situation, check the availability of your load balancer:

```
kubectl get svc
```

The following command output shows that an external IP and port have been allocated:

```
ubuntu@controlplane:~$ kubectl get svc
NAME                   TYPE           CLUSTER-IP       EXTERNAL-IP    PORT(S)        AGE
kubernetes             ClusterIP      10.152.183.1     <none>         443/TCP        31d
whoami-service         ClusterIP      10.152.183.28    <none>         80/TCP         6m18s
metallb-load-balancer  LoadBalancer   10.152.183.213   192.168.1.11   80:31633/TCP   118s
ubuntu@controlplane:~$
```

Figure 7.31 – Load balancer has allocated external IP and port

Now that an external IP has been allocated, we can access the application from any of the nodes or from an external network, as follows:

```
curl 192.168.1.11/whoami
```

The following command output shows that the command was successfully run and the load balancer is working properly. Though the application is spread throughout the nodes, our `whoami` application is able to effectively serve user requests:

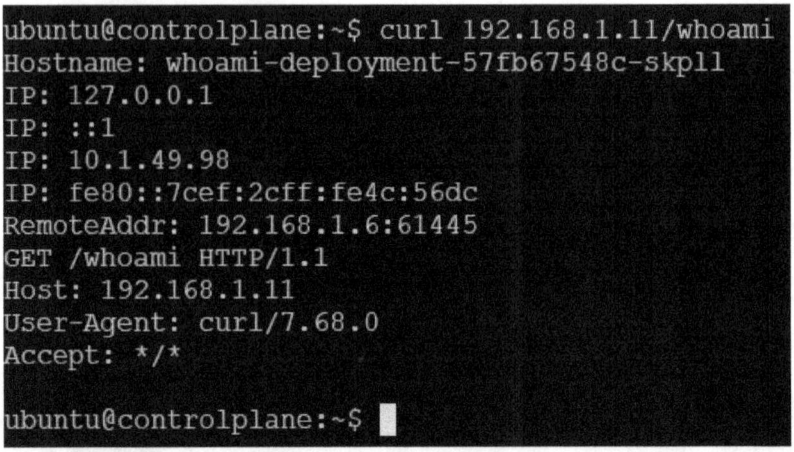

```
ubuntu@controlplane:~$ curl 192.168.1.11/whoami
Hostname: whoami-deployment-57fb67548c-skpll
IP: 127.0.0.1
IP: ::1
IP: 10.1.49.98
IP: fe80::7cef:2cff:fe4c:56dc
RemoteAddr: 192.168.1.6:61445
GET /whoami HTTP/1.1
Host: 192.168.1.11
User-Agent: curl/7.68.0
Accept: */*

ubuntu@controlplane:~$
```

Figure 7.32 – Checking whether the load balancer is working properly

To summarize, Kubernetes provides several ways to expose Services to the outside world. LoadBalancer and Ingress controllers are the most common choices. We have explored both options with examples in this chapter.

Guidelines on how to choose the right load balancer for your applications

Now that we've gone over choices, it would be useful to have a cheat sheet to rapidly compare some crucial features to assist us in deciding which one to utilize.

In the following table, we will cover some of the important parameters in choosing the right option:

Parameters	LoadBalancer	Ingress
Supported by core Kubernetes	Yes	Yes
Works on every platform Kubernetes will deploy	Used in on-premises environments, MetalLB is more suited to this kind of scenario	Yes
Direct access to Service (that is, each Service will get its own external IP)	Yes	No

Parameters	LoadBalancer	Ingress
Proxies each Service through a third party (nginx, HAProxy, and so on)	No	Yes
Multiple ports per Service	Yes	Yes
Multiple Services per IP	No	Yes
Allows use of standard Service ports (80, 443, and so on)	Yes	Yes
Must track individual node IPs	No	No

Table 7.1 – How to choose the right load balancer

In conclusion, it all boils down to a few choices. When using LoadBalancer, especially on bare metal, it works great because the Service can choose which port it wishes to use. The disadvantage is that it can be costly, as each Service will have its own load balancer and external IP address, both of which cost money in the cloud environment. Ingress is becoming the most popular Service when connected with a MetalLB load balancer because it reduces the number of IPs used while still allowing each service to have its own name and/or **Uniform Resource Identifier (URI)** routing.

Summary

In this chapter, we looked at techniques for exposing Services outside the cluster and we've seen how load balancers can expose applications to the outside network. Incoming requests are routed to your application using a load balancer's single IP address. MetalLB implements the LoadBalancer Kubernetes service. When a LoadBalancer Service is requested, MetalLB assigns an IP address from a preset range to the client and informs the network that the IP resides in the cluster.

We have also seen the NGINX Ingress Controller option, which is a common Kubernetes Ingress option. MetalLB, which can be deployed in the same Kubernetes cluster along with Ingress, can also be used as a load balancer. NodePort is another way to expose the Ingress controller to the outside world. Both options were discussed in this chapter, along with different examples.

In the next chapter, we will be covering how to monitor the health of infrastructure and applications using tools such as Prometheus, Grafana, Elastic, Fluentd, Kibana, and Jaeger. You will also learn how to configure and access the various dashboards/metrics.

## 8

Monitoring the Health of Infrastructure and Applications

In the last chapter, we looked at how to expose Services outside of the cluster, and load balancers were used to expose applications to the outside network. The single **Internet Protocol (IP)** address of a load balancer is used to redirect incoming requests to your application. The `LoadBalancer` Kubernetes service is implemented by `MetalLB`. `MetalLB` assigns a client an IP address from a predefined range when a `LoadBalancer` service is requested and informs the network that the IP address is in the cluster. `MetalLB`, which may be deployed alongside Ingress in the same Kubernetes cluster, can also be utilized as a load balancer. Another technique to expose the Ingress controller to the outside world is through `NodePort`. Both options were explored in detail in the previous chapter, with various examples.

In this chapter, we will look at various options for monitoring, logging, and alerting for your infrastructure and applications. In a traditional, host-centric infrastructure, there used to be only two levels of monitoring: applications and the hosts that run them. Then, container abstraction came in, between the host and your applications, after which Kubernetes came in to orchestrate your containers.

To manage infrastructure thoroughly, Kubernetes must now be observed as well. As a result, four distinct components must now be monitored, each with its own set of challenges:

- Hosts
- Containers
- Applications
- And finally, the Kubernetes cluster itself

To keep track of the health of the Kubernetes infrastructure, there is a need to collect metrics and events from all containers and Pods. However, to fully comprehend what clients or users are going through, there is now a need to keep track of the applications that are operating in these Pods. Note that you normally have very little influence over where workloads run when using Kubernetes, which automatically schedules them.

When it comes to monitoring, Kubernetes forces you to reconsider your strategy. But if you know what to look for, where to look for it, and how to aggregate and analyze it, you can make sure your applications are running smoothly and Kubernetes is doing its job effectively.

For aggregating and reporting monitoring data from your cluster, the Kubernetes ecosystem currently offers two in-built add-ons, as detailed next.

Metrics Server collects resource consumption statistics from each `kubelet` on each node and returns aggregated metrics via the `Metrics` **application programming interface (API)**. Metrics Server only stores near-real-time metrics in memory, so it's best used for spot checks of the **central processing unit (CPU)** or memory utilization, or for periodic querying through a full-featured monitoring service that keeps data for long periods of time. The `kube-state-metrics` add-on service makes cluster state data public. Unlike Metrics Server, which provides metrics on Pod and node resource utilization, `kube-state-metrics` polls the control plane API server for information on the overall status of Kubernetes objects (nodes, Pods, Deployments, and so on), as well as resource restrictions and allocations. The information is then utilized to generate metrics, which may be accessed through the Metrics API. In short, `kube-state-metrics` focuses on creating whole new metrics from Kubernetes' object state, whereas `metrics-server` merely saves the most recent data and is not responsible for transmitting metrics to third-party destinations.

In the following sections, we'll go over in detail the various options for retrieving metrics using Metrics Server and `kube-state-metrics`. The advantage of MicroK8s is that the monitoring tools can be enabled in under a minute with only a few commands. It is small enough to fit on a Raspberry Pi and it can be used to develop a monitoring stack that can be deployed anywhere, even at the edge. Furthermore, this is built using some of the most popular open source components that come preinstalled with MicroK8s.

We'll look at how to easily deploy monitoring tools at the edge in this chapter. Such a deployment provides privacy, low latency, and minimal bandwidth costs in **internet of things (IoT)**/edge applications. In this chapter, we're going to cover the following main topics:

- Overview of monitoring, logging, and alerting options
- Configuring a monitoring and alerting stack using the Prometheus, Grafana, and Alertmanager tools
- Configuring a monitoring, logging, and alerting stack using the **Elasticsearch, Fluentd, and Kibana (EFK)** toolset
- Key metrics that need to be monitored

Overview of monitoring, logging, and alerting options

Kubernetes has a lot of advantages, but it also adds a lot of complexity. Its capacity to distribute containerized applications across several nodes and even different data centers (cloud providers, for example) necessitates a comprehensive monitoring solution that can collect and aggregate metrics from a variety of sources.

Many free and paid solutions provide real-time monitoring of Kubernetes clusters and the applications they host, and continuous monitoring of system and application health is critical. Here, we list some prominent open source Kubernetes monitoring tools:

1. **Metrics Server** (in-built) collects resource metrics from `kubelets` and exposes them in the Kubernetes API server through the following Metrics API endpoints:

Kubernetes components	Runs on	Metrics endpoints
`kube-apiserver`	All hosts	`/metrics`
`kube-proxy`	All hosts	`/metrics`

`kubelet`	All hosts	`/metrics`
		`/metrics/cadvisor`
		`/metrics/resource`
		`/metrics/probes`

`kube-scheduler`	At most three hosts that are `dqlite` voters (in case of **High Availability** (**HA**) setup)	`/metrics`
`kube-controller-manager`	At most three hosts that are `dqlite` voters (in case of HA setup)	`/metrics`

Table 8.1 – Metrics API endpoints

2. **Kubernetes Dashboard** (in-built) is a web **user interface** (**UI**) add-on for Kubernetes clusters that allows you to keep track of workload health. Using a simple web interface, Kubernetes Dashboard allows you to manage cluster resources and troubleshoot containerized applications. It provides a concise overview of cluster-wide and individual node resources. It also lists all clusters' namespaces as well as all storage classes that have been declared.

3. **Prometheus** is an open source system for collecting metrics on Kubernetes health. It deploys node exporter Pods on each cluster node, and its server collects data from nodes, Pods, and jobs. Final time-series metrics data is saved in a database, and alerts can be generated automatically based on predefined conditions.

    Prometheus has its own dashboard with limited capabilities that have been extended by the usage of other visualization tools such as Grafana, which uses the Prometheus database to provide advanced inquiries, debugging, and reporting designed for development, test, and production teams.

    It was built with the objective of monitoring applications and microservices in containers at scale, and it can connect to a wide range of third-party databases and supports the bridging of data from other tools. It is made up of three components at its core, as outlined here:

    *   All metrics data will be stored in an in-built time-series database.

    *   A data retrieval worker is in charge of obtaining metrics from outside sources and entering them into the database.

    *   A web server with a simple web interface for configuring and querying the stored data.

    Some of the key features of Prometheus are presented here:

    *   Time-series data classified by metric name and key/value pairs in a multidimensional data model.

    *   Using **Prometheus Query Language** (**PromQL**), a flexible query language that allows us to make use of this dimensionality without relying on distributed storage.

    *   Single-server nodes are self-contained, and time series are collected using a pull model over **HyperText Transfer Protocol** (**HTTP**).

    *   Alternatively, an intermediary gateway can be used to push time series to destinations that are discovered using service discovery or static configuration.

    *   Multiple graphing and dashboarding options are supported.

4. **Grafana**, an open source analytics and metric visualization platform, includes four dashboards: **Cluster**, **Node**, **Pod/Container**, and **Deployment**. Grafana and the Prometheus data source are frequently used by Kubernetes administrators to create information-rich dashboards.

5. **Elasticsearch, Fluentd, and Kibana** make up the EFK stack, which is a combination of three tools that function well together. Fluentd is a data collector for Kubernetes cluster nodes that collects logs from Pods. It sends these logs to the Elasticsearch search engine, which ingests and stores the data in a central location. The EFK stack's UI is Kibana, a data visualization plugin for Elasticsearch that allows users to visualize collected logs and metrics and construct custom dashboards.

Now that we've seen a variety of choices for monitoring, logging, and alerting, we'll go over how to configure them.

Configuring a monitoring and alerting stack using the Prometheus, Grafana, and Alertmanager tools

MicroK8s ships pre-integrated add-ons with Prometheus Operator for Kubernetes, which handles simplified monitoring definitions for Kubernetes services, as well as Prometheus instance deployment and management.

> **Note**
>
> Operators are Kubernetes-specific applications (Pods) that automate the configuration, management, and optimization of other Kubernetes deployments. Operators typically take care of the following:
>
> a. Installing your Kubernetes cluster's specifications and offering initial setup and sizing for your deployment.
>
> b. Reloading Deployments and Pods in real time to accommodate any user-requested parameter changes (hot config reloading).
>
> c. Scaling up or down automatically based on performance data.
>
> d. Backups, integrity checks, and other maintenance tasks should all be performed.

Once the Prometheus add-on is enabled, Prometheus Operator takes care of the installation and configuration of the following items:

- **Kubernetes-Prometheus stack:**

  A. Prometheus servers

  B. Alertmanager

  C. Grafana

  D. Host-node exporter

  E. `kube-state-metrics`

- **ServiceMonitor** entities that define metric endpoint autoconfiguration.

- **Operator Custom Resource Definitions (CRDs) and ConfigMaps** that can be used to customize and scale the services, thus making our configuration entirely portable and declarative.

The following CRDs are managed by Prometheus Operator:

- **PrometheusDeployment**—The Operator ensures that a deployment that matches the resource definition is operating at all times.

- **ServiceMonitor**—Declaratively specifies how to monitor groups of services. Based on the definition, the Operator produces a Prometheus scrape setup automatically.

- **PrometheusRule**—Specifies a Prometheus rule file to be loaded by a Prometheus instance with Prometheus alerting rules.

- **AlertManager**—Specifies the Alertmanager deployment that is desired. The Operator ensures that a deployment that matches the resource definition is operating at all times.

For more information on Prometheus Operator, please refer to the following link:

`https://github.com/prometheus-operator/prometheus-operator`

The following diagram shows the components that we discussed previously:

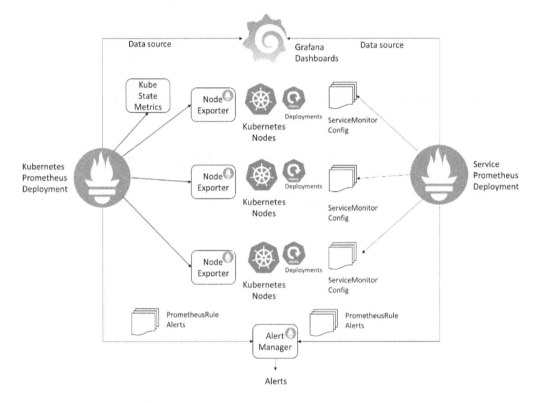

Figure 8.1 – Prometheus Operator components

To summarize, we are going to use the following tools to collect, aggregate, and visualize metrics:

- Kubernetes metrics pulled from Metrics Endpoints.

- Host metrics using Prometheus Node Exporter.

- Alerting using Prometheus Alertmanager.

- Prometheus gathers data from configured targets (from the Kubernetes endpoints discussed in the previous section) at predetermined intervals, analyses rule expressions, displays the results, and can also send out alerts when certain criteria are matched.

- Visualization using Grafana pre-built dashboards.

Now that we are clear on the tools, we will dive into the steps of configuring a monitoring and alerting stack. The following diagram depicts our Raspberry Pi cluster setup:

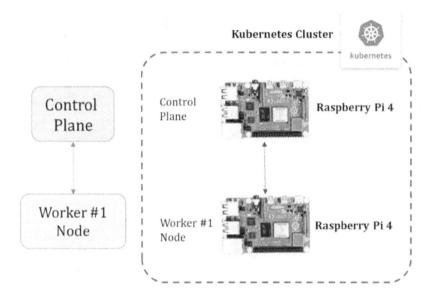

Figure 8.2 – Raspberry Pi cluster setup

Now that we know what we want to do, let's look at the requirements.

Requirements for setting up a MicroK8s Raspberry Pi cluster

Before you begin, here are the prerequisites for building a Raspberry Pi Kubernetes cluster:

- A microSD card (4 **gigabytes** (**GB**) minimum; 8 GB recommended)

- A computer with a microSD card drive

- A Raspberry Pi 2, 3, or 4 (one or more)

- A micro-USB power cable (USB-C for the Pi 4)

- A Wi-Fi network or an Ethernet cable with an internet connection

- (Optional) A monitor with a **High-Definition Multimedia Interface** (**HDMI**) interface

- (Optional) An HDMI cable for the Pi 2 and 3 and a micro-HDMI cable for the Pi 4

- (Optional) A **Universal Serial Bus (USB)** keyboard

Now that we've established the requirements, we'll go on to the step-by-step instructions on how to complete the process.

Step 1 – Creating a MicroK8s Raspberry Pi cluster

Please follow the steps that we covered in *Chapter 5, Creating and Implementing Updates on Multi-Node Raspberry Pi Kubernetes Clusters,* to create a MicroK8s Raspberry Pi cluster. Here's a quick refresher:

- *Step 1*: Installing the **operating system (OS)** image to a **Secure Digital (SD)** card

- *Step 1a:* Configuring Wi-Fi access settings

- *Step 1b*: Configuring remote access settings

- *Step 1c*: Configuring control group settings

- *Step 1d*: Configuring hostname

- *Step 2*: Installing and configuring MicroK8s

- *Step 3*: Adding worker node

A fully functional multi-node Kubernetes cluster should look like the one shown in the following screenshot. To summarize, we have installed MicroK8s on the Raspberry Pi boards and joined multiple Deployments to form the cluster. We have also added nodes to the cluster:

Figure 8.3 – Fully functional MicroK8s Kubernetes cluster

We can now go to the next step of deploying monitoring tools, as we have a fully functional cluster.

By default, none of the MicroK8s add-ons are turned on. As a result, Grafana and Prometheus must be activated post-installation.

Step 2 – Configuring Prometheus, Grafana, and Alertmanager

In this section, we'll enable the Prometheus add-on and access the Prometheus and Grafana dashboards so that we can monitor the Kubernetes cluster and can view alerts if something goes wrong. Use the following command to enable the Dashboard and the Prometheus add-on:

```
microk8s enable dashboard prometheus
```

The following command execution output indicates the Dashboard and the Prometheus add-on have been enabled successfully:

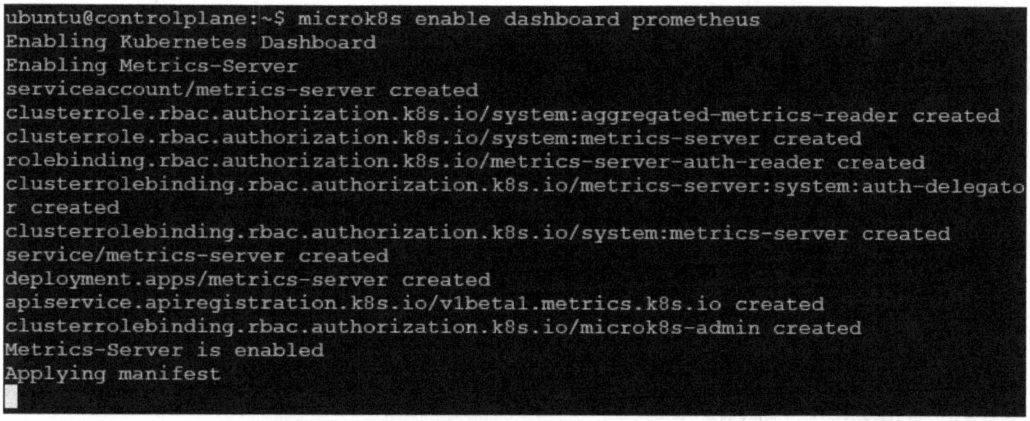

```
ubuntu@controlplane:~$ microk8s enable dashboard prometheus
Enabling Kubernetes Dashboard
Enabling Metrics-Server
serviceaccount/metrics-server created
clusterrole.rbac.authorization.k8s.io/system:aggregated-metrics-reader created
clusterrole.rbac.authorization.k8s.io/system:metrics-server created
rolebinding.rbac.authorization.k8s.io/metrics-server-auth-reader created
clusterrolebinding.rbac.authorization.k8s.io/metrics-server:system:auth-delegato
r created
clusterrolebinding.rbac.authorization.k8s.io/system:metrics-server created
service/metrics-server created
deployment.apps/metrics-server created
apiservice.apiregistration.k8s.io/v1beta1.metrics.k8s.io created
clusterrolebinding.rbac.authorization.k8s.io/microk8s-admin created
Metrics-Server is enabled
Applying manifest
```

Figure 8.4 – Enabling Dashboard and the Prometheus add-on

It will take some time to finish activating the add-on, but the following command execution output shows that Prometheus has been successfully enabled:

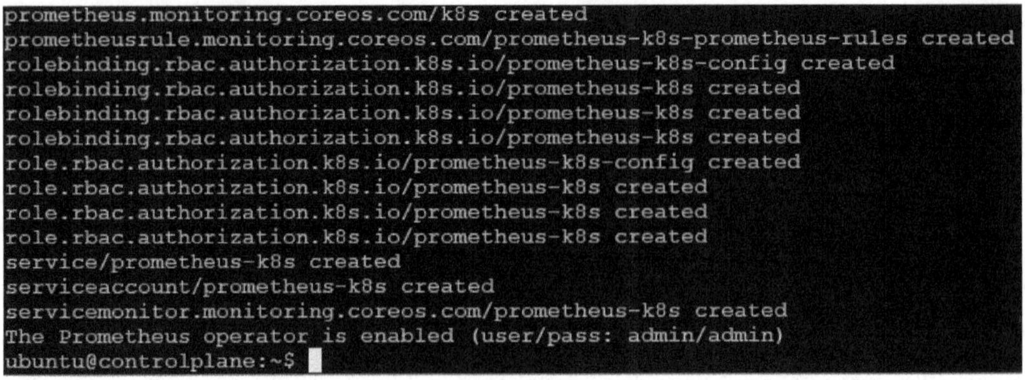

```
prometheus.monitoring.coreos.com/k8s created
prometheusrule.monitoring.coreos.com/prometheus-k8s-prometheus-rules created
rolebinding.rbac.authorization.k8s.io/prometheus-k8s-config created
rolebinding.rbac.authorization.k8s.io/prometheus-k8s created
rolebinding.rbac.authorization.k8s.io/prometheus-k8s created
rolebinding.rbac.authorization.k8s.io/prometheus-k8s created
role.rbac.authorization.k8s.io/prometheus-k8s-config created
role.rbac.authorization.k8s.io/prometheus-k8s created
role.rbac.authorization.k8s.io/prometheus-k8s created
role.rbac.authorization.k8s.io/prometheus-k8s created
service/prometheus-k8s created
serviceaccount/prometheus-k8s created
servicemonitor.monitoring.coreos.com/prometheus-k8s created
The Prometheus operator is enabled (user/pass: admin/admin)
ubuntu@controlplane:~$
```

Figure 8.5 – Add-ons activated

Grafana cannot be enabled with a command. When the Kubernetes Dashboard is enabled, it starts automatically.

To access the Kubernetes Dashboard, we need to create a user and admin role binding. In the next steps, we will create a deployment for it:

```
apiVersion: v1
kind: ServiceAccount
metadata:
  name: admin-user
  namespace: kube-system
---
apiVersion: rbac.authorization.k8s.io/v1
kind: ClusterRoleBinding
metadata:
  name: admin-user
roleRef:
  apiGroup: rbac.authorization.k8s.io
  kind: ClusterRole
  name: cluster-admin
subjects:
- kind: ServiceAccount
  name: admin-user
  namespace: kube-system
```

Create a `dashboard-adminuser.yaml` file with the preceding content and use the following command to create a user and admin role binding:

```
kubectl apply -f dashboard-adminuser.yaml
```

The following command execution output confirms that there is no error in the deployment:

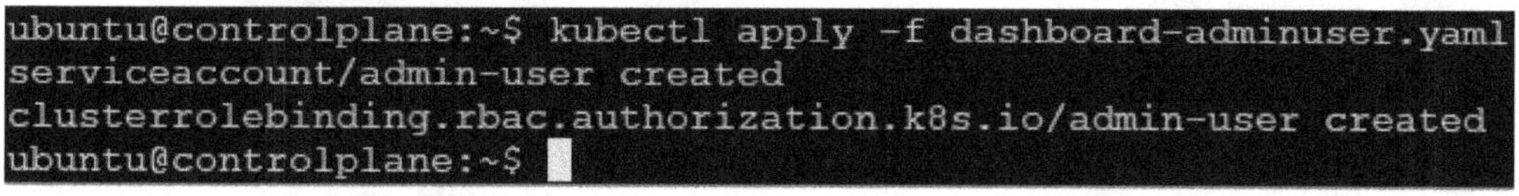

```
ubuntu@controlplane:~$ kubectl apply -f dashboard-adminuser.yaml
serviceaccount/admin-user created
clusterrolebinding.rbac.authorization.k8s.io/admin-user created
ubuntu@controlplane:~$
```

Figure 8.6 – Creating a user and admin role binding

To access the dashboard, we need an access token, which can be obtained by invoking the `kubectl` command, as follows:

```
kubectl -n kube-system describe secret $(microk8s kubectl -n
kube-system get secret | grep admin-user | awk '{print $1}')
```

Copy the token from the command's output and use it in the following step.

It will be necessary to build a secure channel to the cluster with the following command in order to access the Kubernetes Dashboard:

```
kubectl proxy &
```

The following command execution output confirms that a secure channel has been created and we can access the dashboard in the next step:

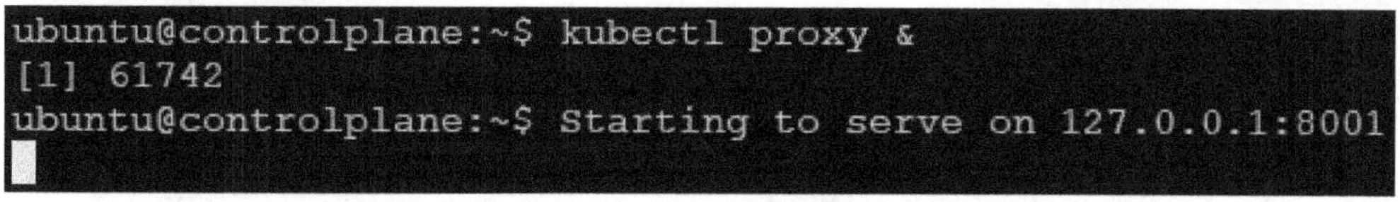

```
ubuntu@controlplane:~$ kubectl proxy &
[1] 61742
ubuntu@controlplane:~$ Starting to serve on 127.0.0.1:8001
```

Figure 8.7 – Creating a secure channel for the dashboard

After that, you'll be able to access the dashboard at the following address:

```
http://<ip address>>:8001/api/v1/namespaces/kube-system/
services/https:kubernetes-dashboard:/proxy/
```

By copying and pasting the token generated in the previous step, you will have access to the cluster's web-based **command-line interface** (**CLI**), as illustrated in the following screenshot:

Figure 8.8 – Kubernetes Dashboard

As discussed earlier, Dashboard is a Kubernetes UI that can be accessed through the web. It can be used to deploy containerized applications to a Kubernetes cluster, troubleshoot them, and control the cluster's resources. The dashboard can be used for a variety of purposes, including the following:

- All nodes and persistent storage volumes are listed in **Admin overview**, along with aggregated metrics for each node.

- **Workloads view** displays a list of all running applications by namespace, as well as current Pod memory utilization and the number of Pods in a Deployment that are currently ready.

- **Discover view** displays a list of services that have been made public and have enabled cluster discovery.

- Drilling down logs from containers belonging to a single Pod is possible using the **Logs viewer** functionality.

- For each clustered application and all Kubernetes resources running in the cluster, **Storage view** identifies persistent volume claims.

We will go to the next step of accessing Prometheus, Grafana, and Alertmanager now that we've enabled all of the required add-ons.

Step 3 – Accessing Prometheus, Grafana, and Alertmanager

We can validate whether Grafana, Prometheus, and Alertmanager are running on the cluster before moving on to other steps.

Navigate to **Monitoring** under **Namespaces** on the Kubernetes Dashboard, and then click **Services**. A list of monitoring services running on the cluster, as well as cluster IP addresses, internal endpoints, and ports, will be displayed, as illustrated in the following screenshot:

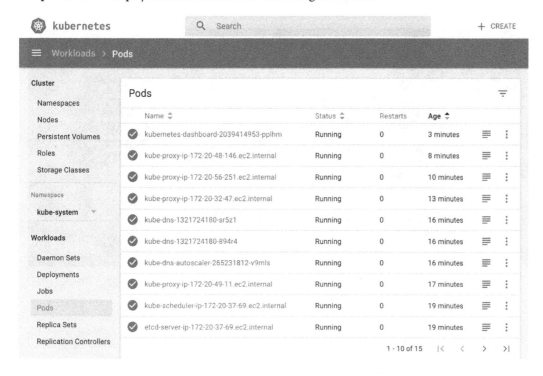

Figure 8.9 – Validating Grafana, Prometheus, and Alertmanager are running on the cluster

From Kubernetes Dashboard, shown in *Figure 8.9*, we can ensure the following components are operational:

- `prometheus-operator` Pod—The core of the stack, in charge of managing other Deployments such as Prometheus servers or Alertmanager servers
- `node-exporter` pod—Per physical host (one in this example)
- `kube-state-metrics` exporter

- **Prometheus server** deployment—`prometheus-k8s` (replicas: 1)

- **Alertmanager** deployment—`alertmanager-main` (replicas: 1)

- **Grafana** deployment—`grafana` (replicas: 1)

The Grafana and Prometheus UIs can then be accessible simply by putting the service IP and ports into the browser in the format `<IP address>:<port>`. The login and password for Grafana will be `admin/admin`.

By default, Grafana uses port `3000`; so, navigate to `http://localhost:3000` in your web browser, and you'll be able to visit the Grafana interface, which is already populated with some interesting dashboards, as we can see here:

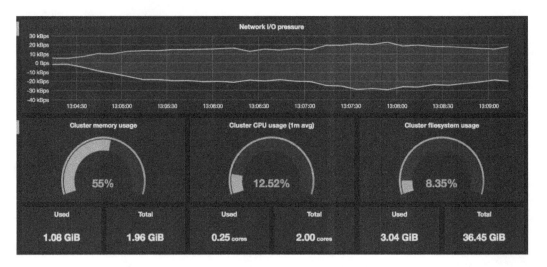

Figure 8.10 – Grafana pre-built dashboards

Grafana comes with Prometheus preinstalled as a data source, as shown in the following screenshot:

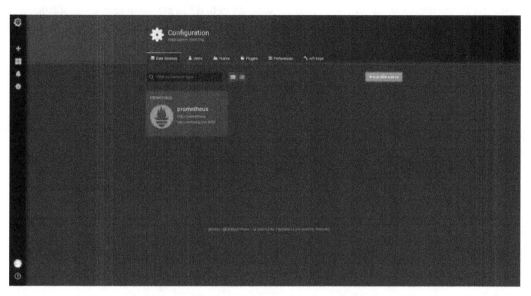

Figure 8.11 – Grafana/Prometheus data source

In a similar way, the Prometheus UI can be accessed. There will be no need for a username or password. By default, Prometheus uses port 9090 and exposes its internal metrics and performance. The Node Exporter Prometheus process runs on port 9100. This exposes the details about the node, including storage space, **random-access memory** (**RAM**), and **central processing unit** (**CPU**) utilization. Metrics should be available on targets with an http://<IP address:9090/metrics path. You can see an overview of the Prometheus UI here:

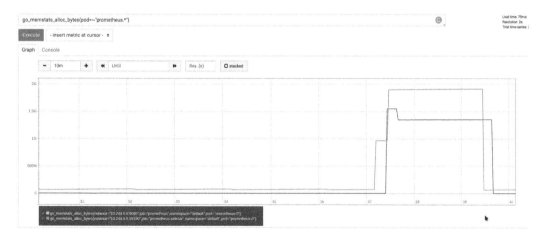

Figure 8.12 – Prometheus UI

Prometheus will scrape and store data based on the predefined configuration. Go to the dashboard to see whether Prometheus has information about the time series that this endpoint exposes on the node.

To see a list of metrics this server is collecting, use the dropdown next to the **Execute** button. A number of metrics prefixed with `node_` that have been collected by Node Exporter can be found in the list. The `cpu metric` node, for example, displays the node's CPU utilization, as illustrated in the following screenshot:

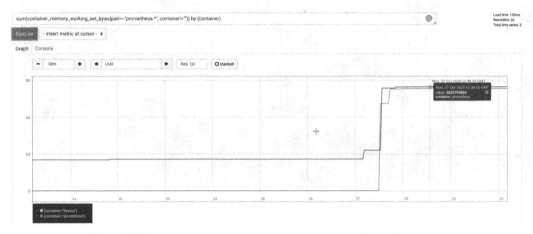

Figure 8.13 – Prometheus metrics visualization

ServiceMonitor automatically detects and registers each target in your Prometheus configuration, as illustrated here:

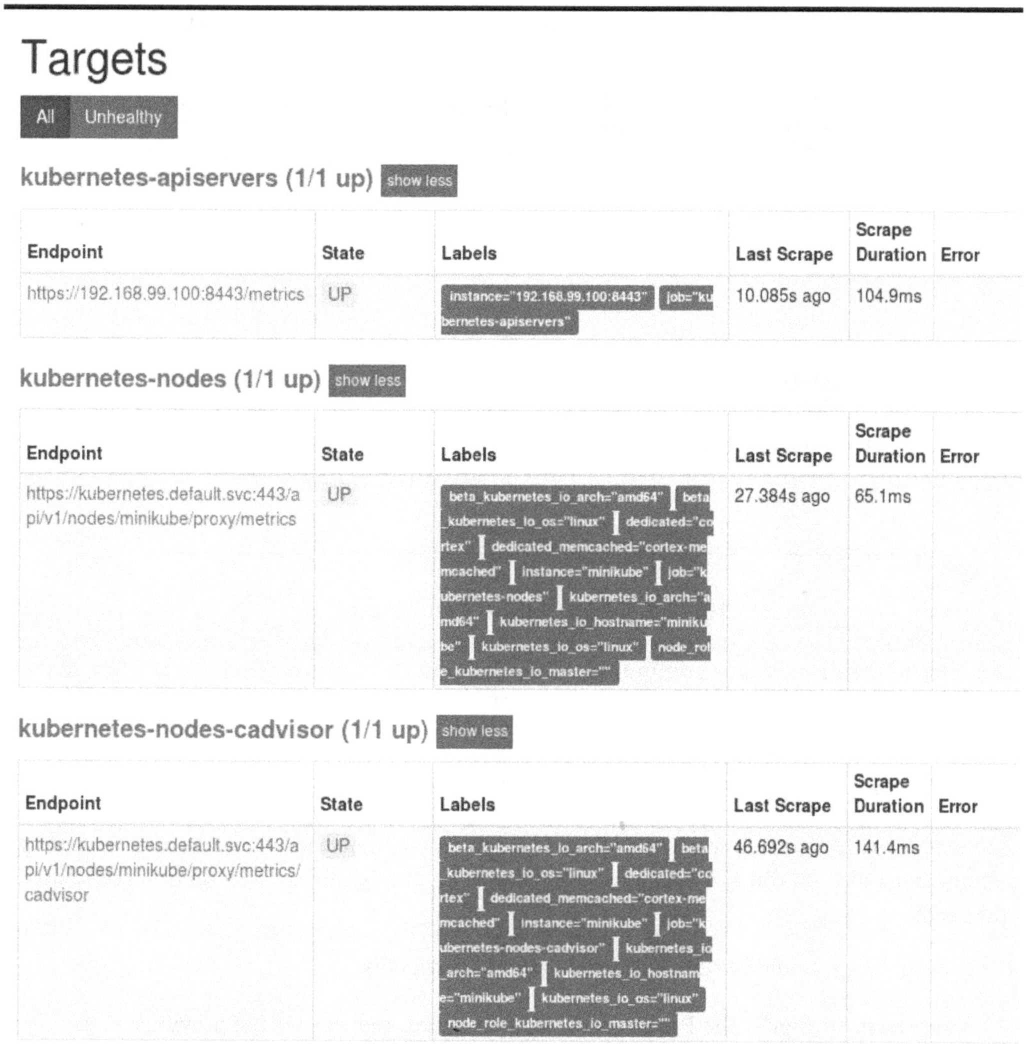

Targets

All   Unhealthy

kubernetes-apiservers (1/1 up) show less

Endpoint	State	Labels	Last Scrape	Scrape Duration	Error
https://192.168.99.100:8443/metrics	UP	instance="192.168.99.100:8443" job="kubernetes-apiservers"	10.085s ago	104.9ms	

kubernetes-nodes (1/1 up) show less

Endpoint	State	Labels	Last Scrape	Scrape Duration	Error
https://kubernetes.default.svc:443/api/v1/nodes/minikube/proxy/metrics	UP	beta_kubernetes_io_arch="amd64" beta_kubernetes_io_os="linux" dedicated="cortex" dedicated_memcached="cortex-memcached" instance="minikube" job="kubernetes-nodes" kubernetes_io_arch="amd64" kubernetes_io_hostname="minikube" kubernetes_io_os="linux" node_role_kubernetes_io_master=""	27.384s ago	65.1ms	

kubernetes-nodes-cadvisor (1/1 up) show less

Endpoint	State	Labels	Last Scrape	Scrape Duration	Error
https://kubernetes.default.svc:443/api/v1/nodes/minikube/proxy/metrics/cadvisor	UP	beta_kubernetes_io_arch="amd64" beta_kubernetes_io_os="linux" dedicated="cortex" dedicated_memcached="cortex-memcached" instance="minikube" job="kubernetes-nodes-cadvisor" kubernetes_io_arch="amd64" kubernetes_io_hostname="minikube" kubernetes_io_os="linux" node_role_kubernetes_io_master=""	46.692s ago	141.4ms	

Figure 8.14 – Prometheus scrape targets

From the Prometheus server interface, the **Alerts** tab displays alerts that are created, as illustrated in the following screenshot:

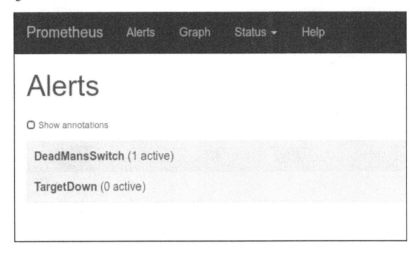

Figure 8.15 – Prometheus Alertmanager

Please refer to the Prometheus community GitHub repository for predefined alert rules, at the following link:

```
https://github.com/prometheus-community/helm-charts/tree/main/
charts/kube-prometheus-stack/templates/prometheus/rules-1.14
```

To summarize, we looked at how to quickly deploy a Kubernetes monitoring and alerting stack using the Prometheus add-on, which is also easy to expand, alter, or migrate to a new set of servers based on the needs.

The following things should be noted for production deployments:

- **Long-term storage**—The Prometheus database stores metrics for the previous 15 days by default. Prometheus doesn't offer long-term storage of metrics. There is no option for backup, data redundancy, trend analysis, data mining, and so on.

- **Authorization and authentication**—There is no server-side authentication, authorization, or encryption provided by Prometheus or its components.

- There is no support for vertical/horizontal scalability.

We've looked at how to use the Prometheus add-on to enable Kubernetes monitoring and alerting, and now, we'll look at how to use the EFK toolset to configure a logging, monitoring, and alerting stack.

Configuring a logging, monitoring, and alerting stack using the EFK toolset

In cases where we need to analyze massive volumes of log data collected by Pods running many services and applications on a Kubernetes cluster, a centralized, cluster-level logging stack could be useful. EFK is the most popular centralized logging solution. Elasticsearch is a real-time search engine that supports full-text and structured searches, as well as analytics, and is distributed and scalable. It's most commonly used for indexing and searching large amounts of log data. Elasticsearch is widely used in conjunction with Kibana, a powerful data visualization frontend and dashboard for Elasticsearch. Kibana is a web-based tool that allows you to quickly query and get insight into Kubernetes applications by viewing Elasticsearch log data and creating dashboards and queries. To gather, transform, and transfer log data to the Elasticsearch backend, we'll use Fluentd, a popular open source data collector, to tail container log files, filter and change data, and feed it to an Elasticsearch cluster for indexing and storage on our Kubernetes nodes. The following diagram depicts what we want to achieve using the EFK toolset:

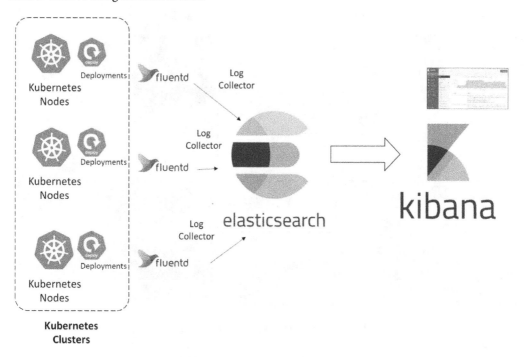

Figure 8.16 – Centralized logging solution: EFK toolset

Since EFK isn't available for `arm64` architecture, I'll be using an Ubuntu **virtual machine** (**VM**) for this section. The instructions for setting up a MicroK8s cluster are the same as in *Chapter 5, Creating and Implementing Updates on Multi-Node Raspberry Pi Kubernetes Clusters*.

Now that we are clear on what we want to achieve, we will dive into the steps in detail.

Step 1 – Enabling the Fluentd add-on

We'll enable the Fluentd add-on in this section, which allows the EFK toolset to gather log data, pass it to Elasticsearch for indexing, and then view aggregated logs using the Kibana dashboard. Use the following command to enable the Fluentd add-on:

```
microk8s enable fluent
```

When you enable this add-on, Elasticsearch, Fluentd, and Kibana (the EFK stack) will be added to MicroK8s.

The following command execution output confirms that the EFK add-on has been enabled:

```
$ microk8s enable fluentd
Enabling Fluentd-Elasticsearch
Labeling nodes
node/host01 labeled
Addon dns is already enabled.
--allow-privileged=true
service/elasticsearch-logging created
serviceaccount/elasticsearch-logging created
clusterrole.rbac.authorization.k8s.io/elasticsearch-logging created
clusterrolebinding.rbac.authorization.k8s.io/elasticsearch-logging created
statefulset.apps/elasticsearch-logging created
configmap/fluentd-es-config-v0.2.0 created
serviceaccount/fluentd-es created
clusterrole.rbac.authorization.k8s.io/fluentd-es created
clusterrolebinding.rbac.authorization.k8s.io/fluentd-es created
daemonset.apps/fluentd-es-v3.1.0 created
deployment.apps/kibana-logging created
service/kibana-logging created
Fluentd-Elasticsearch is enabled
$
```

Figure 8.17 – Enabling Fluentd add-on

Before moving to the next step, let's verify the add-on has been activated.

To do this, use the `microk8s status` command. The following command execution output indicates that the Fluentd add-on has been enabled:

```
$ microk8s status
microk8s is running
high-availability: no
  datastore master nodes: 127.0.0.1:19001
  datastore standby nodes: none
addons:
  enabled:
    dashboard                  # The Kubernetes dashboard
    dns                        # CoreDNS
    fluentd                    # Elasticsearch-Fluentd-Kibana logging and monitoring
    ha-cluster                 # Configure high availability on the current node
    metrics-server             # K8s Metrics Server for API access to service metrics
  disabled:
    ambassador                 # Ambassador API Gateway and Ingress
```

Figure 8.18 – Validating whether the add-on is activated

All of the services for EFK are active, as shown in the output of the command shown next:

```
$ microk8s kubectl get svc -n kube-system
NAME                          TYPE        CLUSTER-IP       EXTERNAL-IP   PORT(S)                  AGE
kube-dns                      ClusterIP   10.152.183.10    <none>        53/UDP,53/TCP,9153/TCP   6m8s
metrics-server                ClusterIP   10.152.183.109   <none>        443/TCP                  5m14s
kubernetes-dashboard          ClusterIP   10.152.183.129   <none>        443/TCP                  4m56s
dashboard-metrics-scraper     ClusterIP   10.152.183.151   <none>        8000/TCP                 4m55s
elasticsearch-logging         ClusterIP   None             <none>        9200/TCP,9300/TCP        2m7s
kibana-logging                ClusterIP   10.152.183.100   <none>        5601/TCP                 2m5s
$
```

Figure 8.19 – Verifying EFK Pods are running

We now have all the services of EFK up and running. To access the Kibana dashboard, we will need to build a secure channel (as we did for Kubernetes Dashboard) to the cluster with the command shown next:

```
microk8s kubectl port-forward -n kube-system service/kibana-
logging 8181:5601
```

The following command execution output confirms that port forwarding is successful:

```
$ microk8s kubectl port-forward -n kube-system service/kibana-logging 8181:5601
Forwarding from 127.0.0.1:8181 -> 5601
Forwarding from [::1]:8181 -> 5601
```

Figure 8.20 – Creating a secure channel for the Kibana dashboard

The Kibana dashboard should be now available at the following address:

```
http://<IP address>:8001/api/v1/namespaces/kube-system/
services/kibana-logging/proxy/app/kibana
```

To summarize, we now have a completely functional EFK stack that can be configured. The next step is to start defining an index pattern in the Kibana dashboard.

Step 2 – Defining an index pattern

We are going to analyze whether the EFK container can start up logs itself. To do that, we'll need to establish an index pattern. A collection of documents with similar characteristics is referred to as an index. An index is given a name, which is used to refer to it while conducting indexing, searching, updating, and deleting operations on the documents it contains.

Launch the Kibana dashboard, and you should see Kibana welcome page, as shown in the following screenshot:

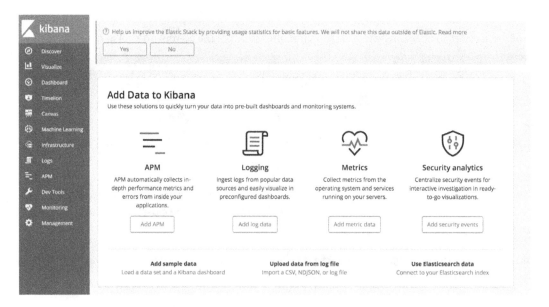

Figure 8.21 – Kibana welcome page

Click **Set up index patterns** on the welcome page or select **Discover item** from the left-hand drop-down menu. On the top, click the **Create index pattern** button. Enter the index's name (for example, `logstash-*`), as shown in the following screenshot:

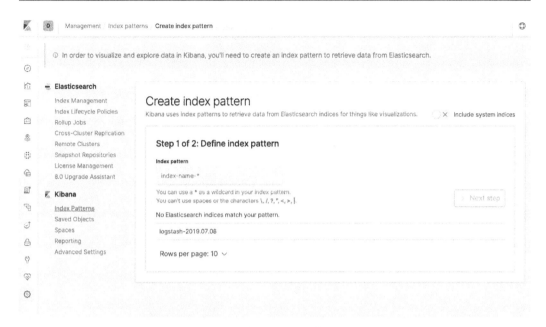

Figure 8.22 – Creating an index pattern

Kibana will then request a field with a time/timestamp that it can use to visualize time-series data. This is the `@timestamp` field in our case, as shown in the following screenshot:

Figure 8.23 – Creating an index pattern with a timestamp field

Click **Create index pattern**, and it should just take a few minutes now that we've built the index pattern. You can see the output here:

Figure 8.24 – Index pattern created

Select the **Discover** option from the left-hand drop-down menu. You should see container log events displayed, as shown in the following screenshot:

Figure 8.25 – Discovering data using the index pattern

The next step is to filter and examine the container startup log events now that we've created an index pattern and organized the data.

Step 3 – Filtering and viewing the data

There will be a listing of all log events with fields available for filtering on the left-hand side, as illustrated in the following screenshot. You may either create a new filter or utilize the `kubernetes.podname` parameter in **Kibana Query Language (KQL)** to filter events:

Figure 8.26 – Filtering log events for a particular Pod

The log list is now filtered to show only log events from that particular Pod. You can explore any event or filter to see more information.

When the Fluent Bit log processor is enabled, it will read, parse, and filter the logs of every Pod on the Kubernetes cluster, enriching each entry with the following data:

- Pod name
- Pod **identifier (ID)**
- Container name
- Container ID
- Labels
- Annotations

Once all the events are indexed, the alerting configuration of Kibana could be used to create rules that detect failure scenarios and then act when those criteria are fulfilled.

More details on alerting can be found here:

```
https://www.elastic.co/guide/en/kibana/current/alerting-getting-
started.html
```

`Fluentd` has a lighter version, Fluent Bit, which was created by the same team for situations with more limited resources. Functionality-wise, `Fluentd` is a log aggregator, while Fluent Bit is just a forwarder. `Fluentd` offers a more robust ecosystem, whereas Fluent Bit is more prevalent in IoT devices.

More details on Fluent Bit can be found here:

```
https://fluentbit.io/
```

Congratulations! Using the EFK stack, we have learned how to aggregate all Kubernetes container logs and analyze them centrally.

To recap, we looked at some of the most popular monitoring, logging, and alerting stack options. The next step is to determine which critical metrics should be monitored in order to manage your infrastructure and applications effectively.

Key metrics that need to be monitored

The rapid adoption of containers in enterprise organizations has provided numerous benefits to developers. However, the flexibility and scalability that Kubernetes provides in deploying containerized applications have also introduced new complications. Keeping track of the health of applications abstracted by containers and then abstracted again by Kubernetes can be difficult without the right tools because there is no longer a 1-to-1 correlation between an application and the server it runs on.

Containerized applications can be spread across multiple environments, and Kubernetes is a complicated environment. Monitoring tools should have the capability to collect metrics from across a distributed environment and deal with the transient nature of containerized resources. Monitoring tools rely on services as their endpoint because Pods and their containers are in constant motion and dynamically scheduled. Services broadcast an IP address that can be accessed from outside Pods, allowing services to communicate in real time as Pods and containers are built and removed.

In Kubernetes, there are two levels of monitoring, as outlined here:

- **Cluster monitoring**—Monitors the health of a Kubernetes cluster as a whole. Helps in checking whether nodes are up to date and running, how many applications are running on each node, and how the cluster as a whole is using resources.

- **Pod monitoring**—Keeps track of issues that affect individual Pods, including a Pod's resource use, application metrics, and metrics linked to replication or autoscaling.

As we discussed in the preceding sections, Kubernetes-based architecture already provides a framework for analyzing and monitoring your applications. You can get a comprehensive view of application health and performance with a suitable monitoring solution that integrates with Kubernetes' built-in abstractions, even if the containers that execute those applications are continually shifting across hosts or scaling up and down.

Next, we will look at some of the key metrics that should be monitored. These are listed here.

Cluster level—The following cluster-state metrics can give you a high-level picture of your cluster's current state. They can reveal problems with nodes or Pods, alerting you to the risk of a bottleneck or the need to expand up your cluster:

What to monitor	Metrics to monitor	Alert criteria
Cluster state	Monitor the aggregated resources usage across all nodes in your cluster, such as the following: Node status Desired Pods Current Pods Available Pods Unavailable Pods	Node status Desired versus current Pods Available and unavailable Pods

Table 8.2 – Cluster-state metrics

Node level—The following measures give you a high-level picture of a node's health and whether or not the scheduler can schedule Pods on it. When you compare resource utilization to resource requests and limits, you may get a better idea of whether your cluster has enough resources to handle its workloads and accommodate new ones. It's critical to maintain and track resource utilization across your cluster's levels, especially for your nodes and the Pods that run on them:

What to monitor	Metrics to monitor	Alert criteria
Node resources	For each node, monitor the following: Memory requests Memory limits Allocatable memory Memory utilization CPU requests CPU limits Allocatable CPU CPU utilization Disk utilization	If the node's CPU or memory usage drops below the desired threshold, look at the following possible causes: Memory limits per Pod versus memory utilization per Pod Memory utilization Memory requests per node versus allocatable memory per node Disk utilization CPU requests per node versus allocatable CPU per node CPU limits per Pod versus CPU utilization per Pod CPU utilization

Table 8.3 – Node-level metrics

Pod level—Although a Pod may be functioning, if it is not accessible, this means it is not ready to accept traffic. This is normal in some situations, such as when a Pod is first launched or when a change to the Pod's specifications is made and deployed. However, if you notice a surge in the number of unavailable Pods or Pods that are constantly unavailable, it could suggest a setup issue. Keep track of the following metrics to gauge the health of Pods:

What to monitor	Metrics to monitor	Alert criteria
Missing Pod	Health and availability of your Pod deployments, such as the following: Available Pods Unavailable Pods	If the number of available Pods for a deployment falls below the number of Pods you specified when you created the deployment
Pods that are not running	If a Pod isn't running or even scheduled, there could be an issue with either the Pod or the cluster, or with your entire Kubernetes deployment, so check the following: Pod status	Alerts should be based on the status of your Pods (Failed, Pending, or Unknown for the period of time you specify)

Table 8.4 – Pod-level metrics

Container level—Some of the container metrics that should be tracked to assess container health are listed next:

What to monitor	Metrics to monitor	Alert criteria
Container restarts	Container restarts could happen when you're hitting a memory limit (for example, **Out-of-Memory (OOM)** kills) in your containers. Also, there could be an issue with either the container itself or its host.	Kubernetes automatically restarts containers, but setting up an alert will give you an immediate notification that you can analyze later and set proper limits.
Container resource usage	Monitor container resource usage for containers in case you're hitting resource limits or spikes in resource consumption	Alerts to check whether container CPU and memory usage and limits are based on thresholds

Table 8.5 – Container metrics

Storage—Volumes serve as a crucial abstraction in the Kubernetes storage architecture. Containers can request storage resources dynamically via a mechanism called volume claims, and volumes can be persistent or non-persistent, as detailed here:

What to monitor	Metrics to monitor	Alert criteria
Storage volumes	Monitor storage to do the following: Ensure your application has enough disk space so that Pods don't run out of space Monitor volume usage and adjust either the amount of data generated by the application or the size of the volume according to usage	Alerts to check whether available bytes capacity exceeds your thresholds. Identify persistent volumes and apply a different alert threshold or notification for these volumes, which likely hold important application data.

Table 8.6 – Storage metrics

Control plane—The worker nodes and Pods in a cluster are managed by the control plane. Here are the control plane components that need to be monitored:

- **etcd**—Stores configuration information that each node in the cluster can use
- **API server**—Validates and configures data for API objects such as Pods, services, and replication controllers, among other things
- **Scheduler**—Manages the use of workloads and the assignment of Pods to available nodes
- **Controller Manager**—A daemon in charge of gathering and sending data to the API server.

You can see more details on these components here:

What to monitor	Metrics to monitor	Alert criteria
etcd	Monitor `etcd` for the following parameters: Leader existence and change rate Committed, applied, pending, and failed proposals **Google Remote Procedure Call (gRPC)** performance	Alerts to check any pending or failed proposals or whether any are reaching inappropriate thresholds
API server	Monitor the API server for the following parameters: Rate/number of HTTP requests Rate/number of `apiserver` requests	Alerts to check whether the rate or number of HTTP requests crosses a desired threshold
Scheduler	Monitor the scheduler for the following parameters: Rate, number, and latency of HTTP requests Scheduling latency Scheduling attempts by result **End-to-end (E2E)** scheduling latency (sum of scheduling)	Alerts to check whether the rate or number of HTTP requests crosses the desired threshold
Controller Manager	Monitor the controller manager for the following parameters: The depth of the work queue The number of retries handled by the work queue	Alerts to check whether requests to the work queue exceed a maximum threshold

Table 8.7 – Control-plane metrics

Kubernetes events

Kubernetes events are a resource type in Kubernetes that are created automatically when other resources' states change, errors occur, or other messages need to be broadcast to the system. Here are various types of events that need to be monitored:

1. **Failed events**—While containers are created on a regular basis, the operation can frequently fail; as a result, Kubernetes does not successfully create that container. Failed events are frequently associated with image retrieval errors. These failures could be caused by typos, insufficient permissions, or upstream build failures. Furthermore, nodes can also fail on their own. When these failures occur, applications should fall back to functional, remaining nodes, but some kind of alerting system is required to determine why the failure occurred. Because a failure is a showstopper—that is, your container will not run until it is resolved—you should pay close attention to this event type.

2. **Evicted events**—Certain Pods can consume a disproportionate amount of computing and memory resources in comparison to their respective runtimes. Kubernetes addresses this issue by evicting Pods and reallocating disk, memory, or CPU space elsewhere.

3. **Storage specific events**—Storage within Pods is commonly used by applications and workloads. Volumes provided by respective providers store critical content that is required by application runtimes. Upon creation, Pods mount these volumes, paving the way for successful operation. These events can alert you when storage volumes are behaving strangely. Furthermore, a node may not be in good enough health to mount a volume. These errors, on the other hand, can make it appear as if a Pod is just getting started. However, discovering these events can assist you in resolving the underlying issues caused by faulty volume mounting.

4. **Kubernetes Scheduling events**—These are scheduling events such as a `FailedScheduling` event occurring when the Kubernetes scheduler is unable to find a suitable node.

5. **Node-specific events**—Node events can indicate erratic or unhealthy behavior elsewhere in the system; for example, the `NodeNotReady` event denotes a node that is not ready for Pod scheduling.

For Kubernetes events, metrics and alert criteria that must be monitored are listed here:

What to monitor	Metrics to monitor	Alert criteria
Kubernetes events	You can monitor how Pod creation, deletion, starting, and stopping affects the performance of your infrastructure by collecting events from Kubernetes and the container engine (such as Docker).	Any failure or exception should be alerted

Table 8.8 – Kubernetes events metrics

Summary

To summarize, we've covered key Kubernetes components, as well as the metrics and events that can help you track their health and performance over time. We have also covered how to collect all of the metrics using built-in Kubernetes APIs and utilities, allowing you to gain comprehensive visibility into your container infrastructure and workloads.

We looked at Prometheus, Grafana, and Alertmanager as tools for setting up a monitoring and alerting stack. We've also looked at how to set up a centralized, cluster-level logging stack with the EFK toolset, which can handle massive amounts of log data. Finally, we went over the essential indicators that should be watched in order to successfully manage your infrastructure and apps.

We'll look at how to develop and deploy a machine learning model using the Kubeflow MLOps platform in the next chapter. Kubeflow and MicroK8s deliver reliable and efficient operations as well as infrastructure optimization.

9

Using Kubeflow to Run AI/MLOps Workloads

In the previous chapter, we looked at several logging, monitoring, and alerting options to gain comprehensive visibility into our container infrastructure and workloads. Regarding tools for setting up a monitoring and alerting stack, we looked at Prometheus, Grafana, and Alert Manager. We also looked at how to use the EFK toolset to set up a centralized, cluster-level logging stack that can handle large volumes of log data. Finally, we discussed the key indicators to keep a close eye on so that you can effectively manage your infrastructure and applications.

In this chapter, we will go through the steps for creating a **machine learning** (**ML**) pipeline that will build and deploy a sample ML model using the Kubeflow MLOps platform. ML is an AI subfield. The purpose of ML is to teach computers to learn from the data you provide. Instead of describing the action, the machine will take your code and provide an algorithm that adjusts, depending on samples of expected behavior. A trained model is the code that results from the combination of the algorithm and the learned parameters.

The following is a high-level overview of the stages in a typical ML workflow:

1. Source and prepare relevant data

2. Develop the ML model

3. Train the model, evaluate model accuracy, and tune the model

4. Deploy the trained model

5. Get predictions from the model

6. Monitor the ongoing predictions

7. Manage the models and their versions

These are iterative stages. At any time during the procedure, you may need to rethink and return to a prior phase.

ML pipelines assist in automating the ML workflow and allowing sequence data to be converted and correlated together in a model so that it can be evaluated. It also allows outputs to be generated. The ML pipeline is designed to allow data to go from raw data format to meaningful information. It provides a way for constructing a multi-ML parallel pipeline system to investigate the outputs of various ML algorithms. There are various stages in a pipeline. Each stage of a pipeline receives data that's been processed from the stage before it – for example, the output of a processing unit is fed into the next phase.

Kubeflow is a platform for data scientists to build and experiment with ML pipelines. It allows you to deploy and build ML workflows. You can specify the ML tools required for your workflow using the Kubeflow configuration. The workflow can then be deployed to the cloud, local, and on-premises multiple platforms for testing and production use.

Data scientists and ML engineers use Kubeflow on MicroK8s to quickly prototype, construct, and deploy ML pipelines. Kubeflow makes MLOps more manageable by bridging the gap between AI workloads and Kubernetes.

Furthermore, Kubeflow on MicroK8s is straightforward to set up and configure, as well as lightweight and capable of simulating production conditions for pipeline creation, migration, and deployment.

In this chapter, we're going to cover the following main topics:

- Overview of the ML workflow
- Deploying Kubeflow
- Accessing the Kubeflow dashboard
- Creating a Kubeflow pipeline to build, train, and deploy a sample ML model
- Recommendations – running AL/ML workloads on Kubernetes

Overview of the ML workflow

Kubeflow aims to be your Kubernetes ML toolkit. The ML tools that are required for your workflow can then be specified using the Kubeflow configurations and the workflow can be deployed to various platforms for testing and production use as required.

Let's have a look at the Kubeflow components before we get into the intricacies of ML workflows.

Introduction – Kubeflow and its components

Kubeflow is a system for deploying, scaling, and managing complex systems based on Kubernetes. For data scientists, Kubeflow is the go-to platform for building and testing ML pipelines. It is also for ML developers and operations teams who wish to deploy ML systems in a variety of contexts for development, testing, and production.

Kubeflow is a framework for establishing the components of your ML system on top of Kubernetes, as shown in the following diagram:

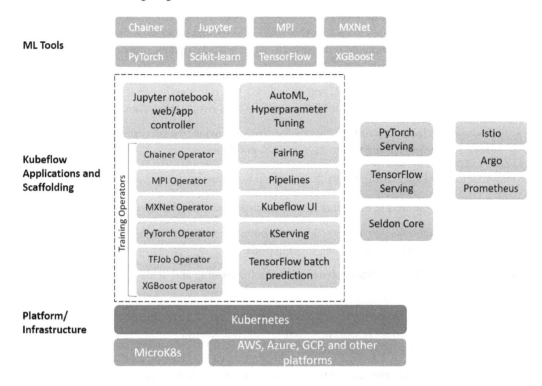

Figure 9.1 – Kubeflow components on top of Kubernetes

We can specify the ML tools required for our workflow by utilizing the Kubeflow configuration interfaces. The workflow can then be deployed to multiple clouds, local, and on-premises platforms for testing and production use.

The following ML tools are supported by Kubeflow:

- **Chainer**: Python-based deep learning framework.

- **Jupyter**: Interactive development environment for notebooks, code, and data that is accessible through the web. Users can create and arrange workflows in data science, scientific computing, computational journalism, and ML using its versatile interface.

- **MPI**: This is a message-passing standard that is standardized and portable and can be used on parallel computing platforms.

- **MXNet**: This is an open source deep learning software framework that's used to train and deploy deep neural networks.

- **PyTorch**: This is an open source ML framework based on the Torch library that's used for applications such as computer vision and natural language processing.

- **Scikit-learn**: This is an ML library for Python that includes support-vector machines, random forests, gradient boosting, k-means, and DBSCAN, among other classification, regression, and clustering techniques, and is designed to work with the Python numerical and scientific libraries known as NumPy and SciPy.

- **TensorFlow**: This is an ML and AI software library that is free and open source. It can be used for a variety of applications, but it focuses on deep neural network training and inference.

- **XGBoost**: This is an open source software library that provides a regularizing gradient boosting framework for C++, Java, Python, R, Julia, Perl, and Scala.

The following are the logical components that make up Kubeflow:

- **Dashboard** allows you to quickly access the Kubeflow components installed in your cluster.

- **Kubeflow Notebooks** allows you to run web-based development environments within your Kubernetes cluster by encapsulating them in pods.

- **Kubeflow Pipelines** is a Docker-based platform for creating and deploying portable, scalable ML workflows.

- **KServing** provides performant, high abstraction interfaces for serving models using standard ML frameworks such as TensorFlow, XGBoost, scikit-learn, PyTorch, and ONNX to tackle production model serving use cases.

- **TensorFlow Serving** takes care of serving functionality for TensorFlow models.

- **PyTorch Serving** takes care of serving functionality for the PyTorch model with Seldon.

- **Seldon** manages, serves, and scales models in any language or framework on Kubernetes.

- **Katib** is a scalable and flexible hyperparameter tuning framework that is tightly integrated with Kubernetes.

- **Training operators** train ML models through operators.

- **Istio Integration** (for TF Serving) provides functionalities such as metrics, auth and quota, rollout and A/B testing, and more.

- **Argo workflows** is a workflow engine that Kubeflow pipelines use to carry out various actions, such as monitoring pod logs, collecting artifacts, managing container life cycles, and more.

- **Prometheus** takes care of logging and monitoring for Kubeflow metrics and Kubernetes components.

- **Multi-Tenancy** is self-served – a new user can self-register to create and own their workspace through the UI. It is currently built around *user namespaces*.

Now that we've seen the various Kubeflow components and ML tools, let's get into the specifics of understanding the ML workflow.

Introduction to the ML workflow

The ML workflow is typically comprised of multiple stages while developing and deploying an ML system. It is an iterative procedure for developing an ML system. To guarantee that the model continues to produce the results you require, you must review the output of various phases of the ML workflow and adjust the model and parameters as needed.

The following diagram shows the experimental phase workflow stages in sequence:

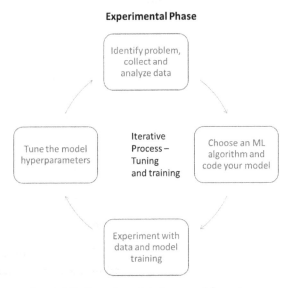

Figure 9.2 – Experimental phase workflow stages

In the experimental phase, the model would be built based on initial assumptions and tested and updated iteratively to achieve the desired results:

1. Determine the problem that needs to be solved by the ML system.
2. Collect and analyze the data required to train the ML model.
3. Select an ML framework and algorithm, and then code the first version of the model.
4. Experiment with the data and model's training.
5. Tune the model's hyperparameters.

The following diagram shows the production phase workflow stages in sequence:

Production Phase

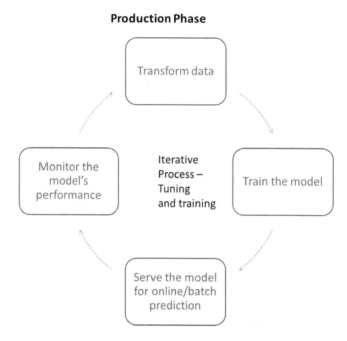

Figure 9.3 – Production phase workflow stages

During the production phase, a system will be deployed that handles the following tasks:

- Transform the data into the format required by the training system. To ensure that the model behaves consistently during training and prediction, the transformation process in the experimental and production phases must be the same.

- Develop the ML model.

- Serve the model for online prediction or batch processing.

- Monitor the model's performance and feed the results into the model for tuning or retraining processes.

Now that we know what all the activities in the experimental and production stages involve, let's look at the Kubeflow components that are involved in each phase.

Kubeflow components in each phase

The following diagram shows the experimental phase workflow stages and the Kubeflow components involved in each stage:

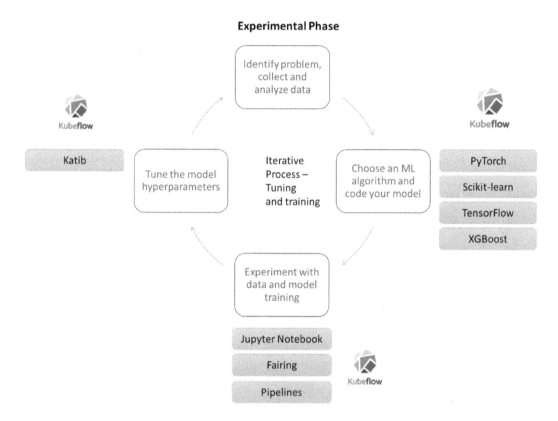

Figure 9.4 – Experimental phase stages and Kubeflow components

The following diagram shows the production phase workflow stages and the Kubeflow components involved in each stage:

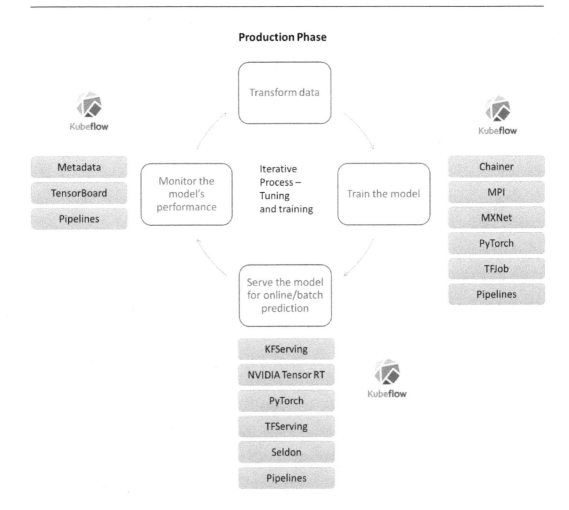

Figure 9.5 – Production phase stages and Kubeflow components

Some of Kubeflow's most important components are as follows:

- Jupyter notebooks can be spawned and managed using Kubeflow's services. Notebooks are used for interactive data science and ML workflow experiments.

- Kubeflow Pipelines is a container-based platform for creating, deploying, and managing multi-step ML processes based on Docker containers.

- Kubeflow has several components that can be used for ML training, hyperparameter tweaking, and workload serving across many platforms.

Now that we've looked at the various Kubeflow components and steps involved in the ML process stages, let's look at Kubeflow Pipelines.

Kubeflow Pipelines

Kubeflow Pipelines are one of the most essential elements of Kubeflow that make AI/ML experiments reproducible, composable, scalable, and easily shareable. Each pipeline component, denoted by a block, is a self-contained piece of code that is packaged as a Docker image. It has inputs (arguments) and outputs, and it completes one stage in the pipeline.

Finally, when the pipeline is run, each container will be executed across the cluster per Kubernetes scheduling, while dependencies are considered. This containerized architecture makes it easy to reuse, share, and swap out components as and when the workflow changes, which is common.

After running the pipeline, the results can be examined in the pipeline's UI on the Kubeflow dashboard. Here, you can debug and tweak parameters and create additional "runs."

To recap, Kubeflow is your ML kit for Kubernetes that offers the following:

- A platform for data scientists who want to build and experiment with ML pipelines
- A platform for ML engineers and operational teams who want to deploy ML systems to various environments for development, testing, and production-level serving
- Services for spawning and managing Jupyter notebooks for interactive data science and experimenting with ML workflows
- A platform for building, deploying, and managing multi-step ML workflows based on Docker containers
- Several components that can be used to build your ML training, hyperparameter tuning, and serving workloads across multiple platforms

In the next section, we'll go over the steps for deploying Kubeflow.

Deploying Kubeflow

Since MicroK8s version 1.22, Kubeflow is no longer available as an add-on; instead, Ubuntu has released Charmed Kubeflow (`https://charmed-kubeflow.io/`), which is a complete collection of Kubernetes operators for delivering the 30+ apps and services that make up the latest version of Kubeflow for easy operations anywhere, from desktops to on-premises, public cloud, and the edge.

Kubeflow is available as a charm, which is a software package that contains a Kubernetes operator as well as information that allows you to integrate many operators into a unified system. This technology utilizes the Juju **Operator Lifecycle Manager** (**OLM**) to provide Kubeflow operations from day 0 to day 2.

To give an overview of Juju, it is an open source modeling tool for cloud-based software operations. It enables you to rapidly and efficiently deploy, set up, manage, maintain, and scale cloud applications on public clouds, as well as physical servers, OpenStack, and containers. More details can be found at `https://ubuntu.com/blog/what-is-juju-introduction-video`.

Juju provides a centralized view of a deployment's Kubernetes operators, including their configuration, scalability, and status, as well as the integration lines that connect them. It keeps track of prospective upgrades and updates for each operator, as well as coordinates the flow of events and communications between them.

Charmed Kubeflow is available in two bundles:

- **Full** (`https://charmhub.io/kubeflow`): Each Kubeflow service is included. At least 14 GB of RAM and 60 GB of storage space is required.

- **Lite** (`https://charmhub.io/kubeflow-lite`): Removes less frequently used services from the complete bundle while maintaining a user-friendly dashboard. This bundle is designed for environments with limited resources.

Now that have a better grasp of Charmed Kubeflow, let's go over the deployment procedure for Kubeflow.

What we are trying to achieve

We wish to do the following in this section:

1. Install and configure Microk8s.
2. Install Juju Operator Lifecycle Manager.
3. Deploy Kubeflow.

Now that we know what we want to do, let's look at the prerequisites for setting up the Kubeflow platform:

- A virtual machine with Ubuntu 20.04 (focal) or later

- At least 16 GB of free memory and 20 GB of disk space

- Access to the internet for downloading the snaps and charms required

Now that we've established the prerequisites, let's learn how to set up the Kubeflow platform.

Step 1 – Installing and configuring MicroK8s

The following steps are similar to the ones we followed in *Chapter 5, Creating and Implementing Updates on Multi-Node Raspberry Pi Kubernetes Clusters* for creating a MicroK8s cluster.

We'll set the snap to install the 1.21 release of Kubernetes since Kubeflow doesn't support the newer 1.22 version yet.

Use the following command to install MicroK8s:

```
sudo snap install microk8s --classic --channel=1.21/stable
```

The following output indicates MicroK8s has been installed successfully:

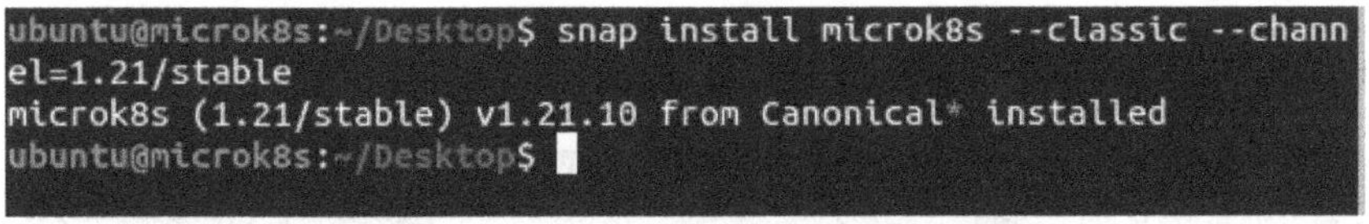

```
ubuntu@microk8s:~/Desktop$ snap install microk8s --classic --chann
el=1.21/stable
microk8s (1.21/stable) v1.21.10 from Canonical* installed
ubuntu@microk8s:~/Desktop$
```

Figure 9.6 – MicroK8s installation

As we saw previously, MicroK8s creates a group called `microk8s` so that it can work without having to use `sudo` for every command. We will be adding the current user to this group to make it easier to run commands:

```
sudo usermod -a -G microk8s $USER
newgrp microk8s
```

Make sure there is proper access to kubectl configuration files as well:

```
sudo chown -f -R $USER ~/.kube
```

As soon as MicroK8s is installed, it will start up. The Kubernetes cluster is now fully operational, as shown in the following screenshot:

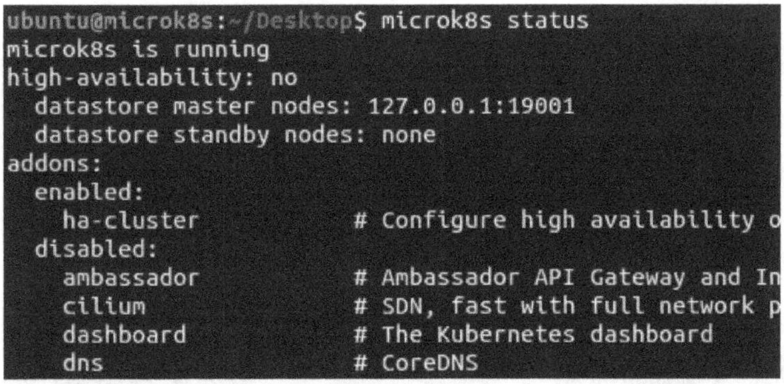

```
ubuntu@microk8s:~/Desktop$ microk8s status
microk8s is running
high-availability: no
    datastore master nodes: 127.0.0.1:19001
    datastore standby nodes: none
addons:
    enabled:
      ha-cluster              # Configure high availability o
    disabled:
      ambassador              # Ambassador API Gateway and In
      cilium                  # SDN, fast with full network p
      dashboard               # The Kubernetes dashboard
      dns                     # CoreDNS
```

Figure 9.7 – MicroK8s is fully operational

Before installing Kubeflow, let's enable some add-ons. We'll set up a DNS service so that the applications can discover one other, as well as storage, an ingress controller for accessing Kubeflow components, and the MetalLB load balancer application. All of these can be enabled at the same time using the following command:

```
microk8s enable dns storage ingress metallb:10.64.140.43-
10.64.140.49
```

With this command, we've instructed MetalLB to give out addresses in the `10.64.140.43 - 10.64.140.49` range.

The following output shows that add-ons are being enabled:

```
ubuntu@microk8s:~/Desktop$ microk8s enable dns storage ingress metallb:10.64.14
0.43-10.64.140.49
Infer repository core for addon dns
Infer repository core for addon storage
Infer repository core for addon ingress
Infer repository core for addon metallb
Addon core/dns is already enabled
Addon core/storage is already enabled
Enabling Ingress
ingressclass.networking.k8s.io/public created
namespace/ingress created
serviceaccount/nginx-ingress-microk8s-serviceaccount created
clusterrole.rbac.authorization.k8s.io/nginx-ingress-microk8s-clusterrole create
d
role.rbac.authorization.k8s.io/nginx-ingress-microk8s-role created
clusterrolebinding.rbac.authorization.k8s.io/nginx-ingress-microk8s created
rolebinding.rbac.authorization.k8s.io/nginx-ingress-microk8s created
```

Figure 9.8 – Add-ons enabled successfully

MicroK8s may take a few minutes to install and configure these extra features. Before we go any further, we should double-check that the add-ons have been correctly activated and that MicroK8s is ready to use. From the following output, we can infer that all the required add-ons are enabled:

```
clusterrole.rbac.authorization.k8s.io/metallb-system:speaker
role.rbac.authorization.k8s.io/config-watcher created
role.rbac.authorization.k8s.io/pod-lister created
clusterrolebinding.rbac.authorization.k8s.io/metallb-system:c
clusterrolebinding.rbac.authorization.k8s.io/metallb-system:s
rolebinding.rbac.authorization.k8s.io/config-watcher created
rolebinding.rbac.authorization.k8s.io/pod-lister created
Warning: spec.template.spec.nodeSelector[beta.kubernetes.io/o
ce v1.14; use "kubernetes.io/os" instead
daemonset.apps/speaker created
deployment.apps/controller created
configmap/config created
MetalLB is enabled
ubuntu@microk8s:~/Desktop$ ▮
```

Figure 9.9 – Add-ons enabled successfully

We can achieve this by using the `microk8s status` command and specifying the `--wait-ready` option, which instructs MicroK8s to complete whatever processes it is currently working on before returning.

Now that we have a running Kubernetes cluster, let's install Juju.

Step 2 – Installing Juju Operator Lifecycle Manager

As we discussed previously, Juju is an OLM for the cloud, bare metal, or Kubernetes. It can be used to deploy and manage the various components that make up Kubeflow.

Like MicroK8s, Juju can be installed from a snap package using the following command:

```
sudo snap install juju --classic
```

The following output shows that the Juju snap has been installed successfully:

Figure 9.10 – Juju installed

No further setup or configuration is required since the Juju OLM recognizes MicroK8s. To deploy a Juju controller to the Kubernetes session we put up with MicroK8s, all we must do is run the following command:

```
juju bootstrap microk8s --agent-version="2.9.22"
```

> **Note**
> In the latest versions, you can just use `latest` instead of specifying the agent version.

The following output shows that the Juju OLM bootstrap configuration has been successful:

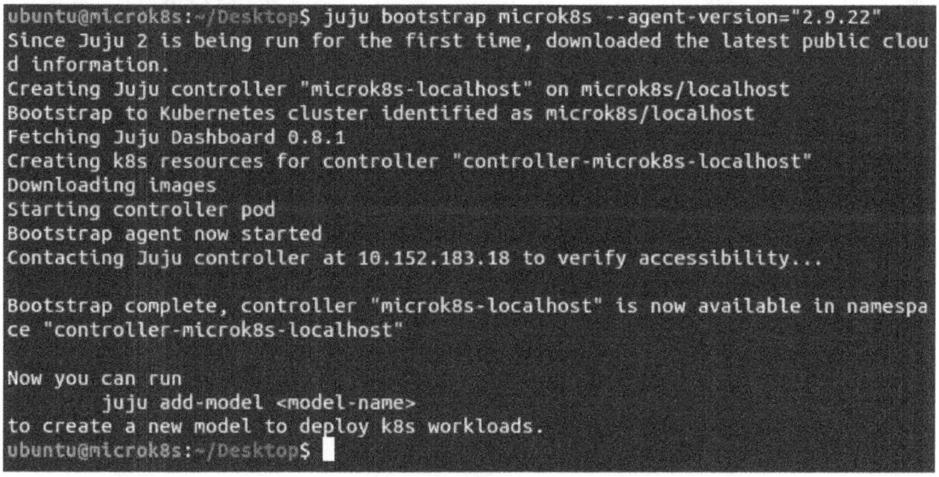

Figure 9.11 – Juju OLM bootstrap

The controller is Juju's Kubernetes-based agent, which may be used to deploy and control Kubeflow components. The controller can work with a variety of models, which correspond to Kubernetes namespaces. Setting up a new model for Kubeflow is the recommended option:

```
juju add-model kubeflow
```

> **Note**
>
> At the time of writing, the model must be named kubeflow, but this is planned to be addressed in future versions.

The following output shows that the kubeflow model was added successfully:

```
ubuntu@microk8s:~/Desktop$ juju add-model kubeflow
Added 'kubeflow' model on microk8s/localhost with credential 'microk8s' for use
r 'admin'
ubuntu@microk8s:~/Desktop$
```

Figure 9.12 – Juju model addition

Now that the Kubeflow model has been added, our next step is to deploy the Charmed Kubeflow bundle. Charmed Kubeflow is essentially a charm collection. Each charm deploys and controls a single application that makes up Kubeflow. You can just install the components that are required by deploying the charms individually and linking them together to create Kubeflow. However, three bundles are offered for your convenience. These bundles are essentially a recipe for a certain Kubeflow deployment, setting and connecting the applications in such a way that you end up with a working deployment with the least amount of effort.

The full Kubeflow bundle will necessitate a lot of resources (at least 4 CPUs, 14 GB of free RAM, and 60 GB of disk space), so starting with the kubeflow-lite bundle is the best alternative:

```
juju deploy kubeflow-lite --trust
```

The following output shows that Juju will start acquiring the apps and start deploying them to the MicroK8s Kubernetes cluster. This procedure can take quite some time but can vary based on your hardware configuration:

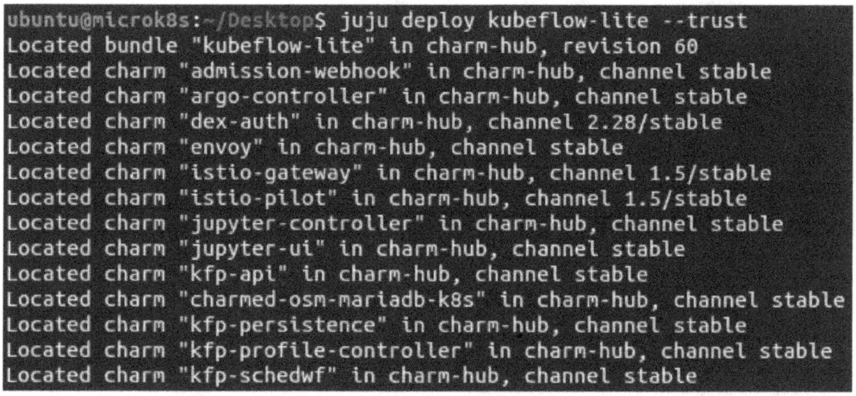

```
ubuntu@microk8s:~/Desktop$ juju deploy kubeflow-lite --trust
Located bundle "kubeflow-lite" in charm-hub, revision 60
Located charm "admission-webhook" in charm-hub, channel stable
Located charm "argo-controller" in charm-hub, channel stable
Located charm "dex-auth" in charm-hub, channel 2.28/stable
Located charm "envoy" in charm-hub, channel stable
Located charm "istio-gateway" in charm-hub, channel 1.5/stable
Located charm "istio-pilot" in charm-hub, channel 1.5/stable
Located charm "jupyter-controller" in charm-hub, channel stable
Located charm "jupyter-ui" in charm-hub, channel stable
Located charm "kfp-api" in charm-hub, channel stable
Located charm "charmed-osm-mariadb-k8s" in charm-hub, channel stable
Located charm "kfp-persistence" in charm-hub, channel stable
Located charm "kfp-profile-controller" in charm-hub, channel stable
Located charm "kfp-schedwf" in charm-hub, channel stable
```

Figure 9.13 – Juju deployment

By running the `juju status` command, you can keep track of the deployment's progress:

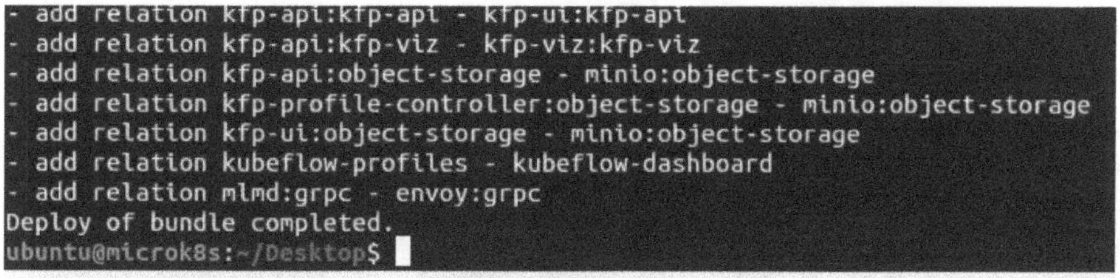

```
- add relation kfp-api:kfp-api - kfp-ui:kfp-api
- add relation kfp-api:kfp-viz - kfp-viz:kfp-viz
- add relation kfp-api:object-storage - minio:object-storage
- add relation kfp-profile-controller:object-storage - minio:object-storage
- add relation kfp-ui:object-storage - minio:object-storage
- add relation kubeflow-profiles - kubeflow-dashboard
- add relation mlmd:grpc - envoy:grpc
Deploy of bundle completed.
ubuntu@microk8s:~/Desktop$
```

Figure 9.14 – Juju deployment successful

There could be error messages since many of the components rely on the operation of others, so it may take some time before everything is up and running.

Now the bundle has been deployed, let's complete some of the post-installation configurations.

Step 3 – Post-installation configurations

Some configuration needs to be set with the URL so that we have authentication and access to the dashboard service. This is dependent on the underlying network provider, but for this local deployment, we know what the URL will be for running on a local MicroK8s. The following commands can be used to configure it in Juju:

```
juju config dex-auth public-url=http://10.64.140.43.nip.io
juju config oidc-gatekeeper public-url=http://10.64.140.43.nip.io
```

The following output confirms that dex-auth public-url and oidc-gatekeeper public-url have been set successfully:

```
ubuntu@microk8s:~/Desktop$ juju config dex-auth public-url
=http://10.64.140.43.nip.io
ubuntu@microk8s:~/Desktop$ juju config oidc-gatekeeper pub
lic-url=http://10.64.140.43.nip.io
```

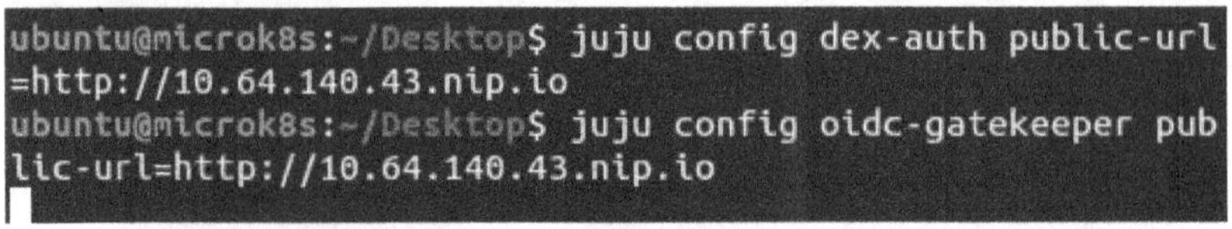

Figure 9.15 – Setting public-url for the dashboard service

Run the following commands to enable basic authentication and create a username and password for the Kubeflow deployment:

```
juju config dex-auth static-username=admin
juju config dex-auth static-password=admin
```

The following output confirms that the username and password of the Kubeflow deployment have been set successfully:

```
ubuntu@microk8s:~/Desktop$ juju config dex-auth static-use
rname=admin
ubuntu@microk8s:~/Desktop$ juju config dex-auth static-pas
sword=admin
```

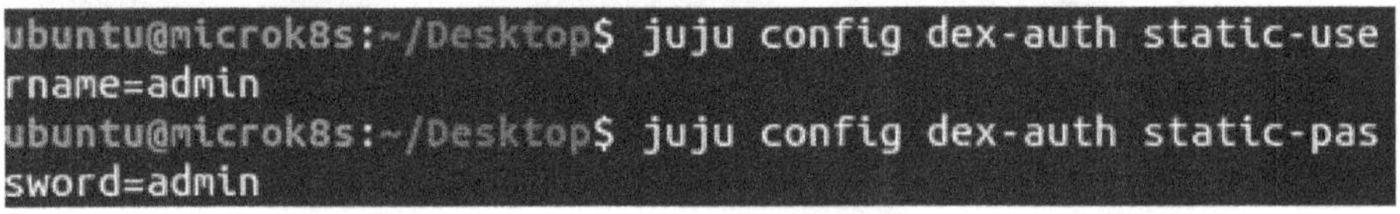

Figure 9.16 – Setting the username and password for the Kubeflow deployment

With that, we have installed and configured MicroK8s. We have also deployed the Juju Charmed Operator Framework to manage apps and automate operations and also integrated all the required Kubeflow components. Finally, we configured the Kubeflow dashboard service's authentication. Now, let's learn how to access the Kubeflow dashboard.

Accessing the Kubeflow dashboard

The Kubeflow dashboard gives you easy access to all the Kubeflow components installed on the cluster. Point your browser to http://10.64.140.43.nip.io (the URL that we set earlier) to be taken to the login screen, where we can input admin as the username and admin as the password (we set these components up previously).

The **Welcome** page should appear. Clicking **Start Setup** will lead you to the **Create namespace** screen. When you enter the namespace and click the **Finish** button, the dashboard will appear, as shown in the following screenshot:

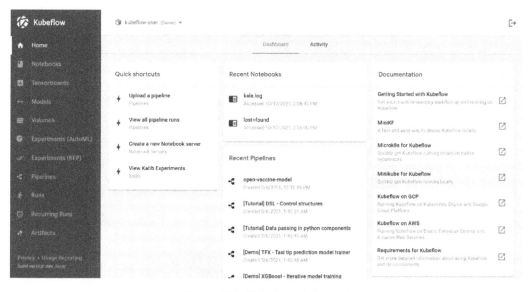

Figure 9.17 – Kubeflow dashboard

Great! We have just installed Kubeflow.

Now that Kubeflow has been installed and is operational, let's learn how to translate an ML model into a Kubeflow pipeline.

Creating a Kubeflow pipeline to build, train, and deploy a sample ML model

In this section, we will be using the Fashion MNIST dataset and TensorFlow's Basic classification to build the pipeline step by step and turn the example ML model into a Kubeflow pipeline.

Before deploying Kubeflow, we will look at the dataset that we are going to use. Fashion-MNIST (https://github.com/zalandoresearch/fashion-mnist) is a Zalando article image dataset that includes a training set of 60,000 samples and a test set of 10,000 examples. Each sample is a 28 x 28 grayscale image with a label from one of 10 categories.

Each training or test item in the dataset is assigned to one of the following labels:

Label	Description
0	T-shirt/top
1	Trouser
2	Pullover
3	Dress

Label	Description
4	Coat
5	Sandal
6	Shirt
7	Sneaker
8	Bag
9	Ankle boot

Table 9.1 – Categories in the Fashion MNIST dataset

Now that our dataset is ready, we can launch a new notebook server via the Kubeflow dashboard.

Step 1 – launching a new notebook server from the Kubeflow dashboard

You can start a new notebook by clicking **New Server** on the **Notebook Servers** tab. Select a Docker **Image** for the notebook server and give it a **Name**. Choose the appropriate **CPU**, **RAM**, and **Workspace volume** values and click **Launch**.

To browse the web interface that's been exposed by your server, click **CONNECT**:

Figure 9.18 – Launch new notebook server

Ensure that **Allow access to Kubeflow Pipelines** is enabled on the new notebook so that we can use the Kubeflow Pipelines SDK in the next step:

Figure 9.19 – New notebook configurations

Launch a new terminal from the right-hand menu (**New | Terminal**) while you're in the notebook server. Download the notebook from GitHub in the terminal using the following `git` command:

```
$ git clone https://github.com/manceps/manceps-canonical.git
```

This will clone and open the `KF_Fashion_MNIST` notebook, as shown in the following screenshot:

From Notebook to Kubeflow Pipeline using Fashion MNIST

In this notebook, we will walk you through the steps of converting a machine learning model, which you may already have on a jupyter notebook, into a Kubeflow pipeline. As an example, we will make use of the fashion we will make use of the fashion MNIST dataset and the Basic classification with Tensorflow example.

In this example we use:

- **Kubeflow pipelines** - Kubeflow Pipelines is a machine learning workflow platform that is helping data scientists and ML engineers tackle experimentation and productionization of ML workloads. It allows users to easily orchestrate scalable workloads using an SDK right from the comfort of a Jupyter Notebook.

- **Microk8s** - Microk8s is a service that gives you the ability to spin up a lightweight Kubernetes cluster right on your local machine. It comes with Kubeflow built right in.

Note: This notebook is to be run on a notebook server inside the Kubeflow environment.

Section 1: Data exploration (as in here)

The Fashion MNIST dataset contains 70,000 grayscale images in 10 clothing categories. Each image is 28x28 pixels in size. We chose this dataset to demonstrate the funtionality of Kubeflow Pipelines without introducing too much complexity in the implementation of the ML model.

To familiarize you with the dataset we will do a short exploration. It is always a good idea to understand your data before you begin any kind of analysis.

Figure 9.20 – Fashion MNIST notebook

In **Section 1** of the notebook, we can familiarize ourselves with the dataset that we have; we'll perform a quick analysis by exploring the data. Here's a quick refresher on the dataset:

- There are 60,000 training labels and 10,000 test labels

- Each label corresponds to one of the 10 class names and is a number between 0 and 9

Before starting any form of analysis, it's always a good idea to comprehend the data. In `section 1.4` of the notebook, we will preprocess the data – that is, the data must be normalized so that each value falls between 0 and 1 to successfully train the model.

The following screenshot shows that the values are between 0 and 255:

```
In [ ]: plt.figure()
        plt.imshow(train_images[0])
        plt.colorbar()
        plt.grid(False)
        plt.show()
```

Figure 9.21 – The first image from the dataset

In the next step, we must divide the training and test values by 255 to scale the data. It's critical that the training and testing sets are both preprocessed the same way:

```
In [ ]: train_images = train_images / 255.0

        test_images = test_images / 255.0
```

Figure 9.22 – Dividing the training and test values by 255

The following is the execution output from the Jupyter notebook for the preceding code:

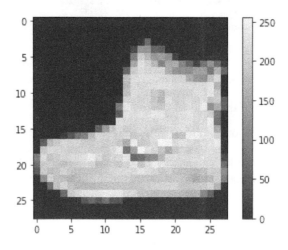

Figure 9.23 – Preprocessing the data

Let's inspect the first 25 images from the training set, together with the class name, to make sure the data is in the right format. Then, we will be ready to build and train the network:

```
In [ ]: plt.figure(figsize=(10,10))
        for i in range(25):
            plt.subplot(5,5,i+1)
            plt.xticks([])
            plt.yticks([])
            plt.grid(False)
            plt.imshow(train_images[i], cmap=plt.cm.binary)
            plt.xlabel(class_names[train_labels[i]])
        plt.show()
```

Figure 9.24 – Plot for the first 25 images from the training set

The following is the output from the Jupyter notebook for the preceding code:

Figure 9.25 – Inspecting 25 images from the training set

Now that the data has been preprocessed, we can build the model.

The TensorFlow model we're working on is an example of basic classification (`https://www.tensorflow.org/tutorials/keras/classification`).

Step 2 – creating a Kubeflow pipeline

We will be using the Kubeflow Pipelines SDK, which is a set of Python packages for specifying and running ML workflows. A pipeline is a description of an ML workflow that includes all the components that make up the steps in the workflow, as well as how they interact.

Install the Kubeflow Pipelines SDK (`kfp`) in the current userspace to ensure that you have access to the necessary packages in your Jupyter notebook instance:

```
In [ ]:  !pip install --user --upgrade kfp
```

Figure 9.26 – Installing the Kubeflow Pipelines SDK

Now that we have installed the Kubeflow Pipelines SDK, the next step is to create Python scripts for Docker containers using `func_to_container_op`.

To package our Python code inside containers, we must create a standard Python function that contains a logical step in your pipeline. In this case, two functions have been defined: `train` and `predict`. Our model will be trained, evaluated, and saved by the train component (refer to `Section 2.2` in the notebook):

```python
# Run a training job with specified number of epochs
model.fit(train_images, train_labels, epochs=10)

# Evaluate the model and print the results
test_loss, test_acc = model.evaluate(test_images, test_labels, verbose=2)
print('Test accuracy:', test_acc)

# Save the model to the designated
model.save(f'{data_path}/{model_file}')

# Save the test_data as a pickle file to be used by the predict component.
with open(f'{data_path}/test_data', 'wb') as f:
    pickle.dump((test_images,test_labels), f)
```

Figure 9.27 – Model trained and saved

The layers of the neural network must be configured before the model can be compiled. The layers are the most fundamental components of a neural network. Data is put into layers, and then representations are extracted from it:

```python
# Define a Softmax layer to define outputs as probabilities
probability_model = tf.keras.Sequential([model,
                                tf.keras.layers.Softmax()])
```

Figure 9.28 – Neural network layers to be configured

The predict component takes the model and applies it to an image from the test dataset to create a prediction:

```python
# See https://github.com/kubeflow/pipelines/issues/2320 for explanation on this line.
image_number = int(image_number)

# Grab an image from the test dataset.
img = test_images[image_number]

# Add the image to a batch where it is the only member.
img = (np.expand_dims(img,0))

# Predict the label of the image.
predictions = probability_model.predict(img)
```

Figure 9.29 – Predicting from the test dataset

Our final step is to convert these functions into container components. The `func_to_container_op` method can be used to accomplish this:

```python
In [ ]:  # Create train and predict lightweight components.
         train_op = comp.func_to_container_op(train, base_image='tensorflow/tensorflow:latest-gpu-py3')
         predict_op = comp.func_to_container_op(predict, base_image='tensorflow/tensorflow:latest-gpu-py3')
```

Figure 9.30 – Converting Python scripts into Docker containers

After converting Python scripts into Docker containers, the next step is to define a Kubeflow pipeline.

Kubeflow makes use of YAML templates to define Kubernetes resources. Without having to manually alter YAML files, the Kubeflow Pipelines SDK allows you to describe how our code is run. It generates a compressed YAML file that defines our pipeline at compile time. This file can then be reused or shared in the future, making the workflow scalable and repeatable.

Our first step is to launch a Kubeflow client, which includes client libraries for the Kubeflow Pipelines API, allowing us to construct more experiments and run within those experiments directly from the Jupyter notebook:

```
In [ ]:  client = kfp.Client(host='pipelines-api.kubeflow.svc.cluster.local:8888')
```

Figure 9.31 – Launching the Kubeflow client

The preceding components build a client to communicate with the pipeline's API server. The next step will be to design the pipeline's various components.

The pipeline function has been defined, and it contains several parameters that will be passed to our various components during execution. Kubeflow Pipelines are built declaratively. This means that the code will not be executed until the pipeline has been compiled:

```
In [ ]:  # Define the pipeline
         @dsl.pipeline(
             name='MNIST Pipeline',
             description='A toy pipeline that performs mnist model training and prediction.'
         )

         # Define parameters to be fed into pipeline
         def mnist_container_pipeline(
             data_path: str,
             model_file: str,
             image_number: int
         ):

             # Define volume to share data between components.
             vop = dsl.VolumeOp(
             name="create_volume",
             resource_name="data-volume",
             size="1Gi",
             modes=dsl.VOLUME_MODE_RWM)

             # Create MNIST training component.
             mnist_training_container = train_op(data_path, model_file) \
                                         .add_pvolumes({data_path: vop.volume})

             # Create MNIST prediction component.
             mnist_predict_container = predict_op(data_path, model_file, image_number) \
                                         .add_pvolumes({data_path: mnist_training_container.pvolume})
```

Figure 9.32 – Defining the pipeline

To save and persist data between components, a Persistent Volume Claim can be quickly created using the VolumeOp method.

VolumeOp's parameters include the following:

- name: The name that's displayed for the volume creation operation
- resource_name: The name that can be referenced by other resources

- `size`: The size of the volume claim

- `modes`: The access mode for the volume

It's finally time to define our pipeline's components and dependencies. This can be accomplished using `ContainerOp`, an object that defines a pipeline component from a container:

```
# Create MNIST training component.
mnist_training_container = train_op(data_path, model_file) \
                            .add_pvolumes({data_path: vop.volume})

# Create MNIST prediction component.
mnist_predict_container = predict_op(data_path, model_file, image_number) \
                            .add_pvolumes({data_path: mnist_training_container.pvolume})
```

Figure 9.33 – Creating the training and prediction components

The `train_op` and `predict_op` components accept arguments from the original Python function. We attach our `VolumeOp` at the end of the function with a dictionary of paths and associated Persistent Volumes to be mounted to the container before execution.

While `train_op` is using the `pvolumes` dictionary's `vop.volume` value, `<Container Op>`, the `pvolume` argument, which is used by the other components, ensures that the volume from the previous `ContainerOp` is used instead of creating a new one:

```
# Print the result of the prediction
mnist_result_container = dsl.ContainerOp(
    name="print_prediction",
    image='library/bash:4.4.23',
    pvolumes={data_path: mnist_predict_container.pvolume},
    arguments=['cat', f'{data_path}/result.txt']
)
```

Figure 9.34 – Attaching a volume to the container

ContainerOp's parameters include the following:

- `name`: The name displayed for the component's execution during runtime

- `image`: The image tag for the Docker container to be used

- `pvolumes`: A dictionary of paths and associated *Persistent Volumes* to be mounted to the container before execution

- `arguments`: The command to be run by the container at runtime

Now that we've created separate components for training and prediction, we can start compiling and running the pipeline code in the notebook.

Step 3 – compiling and running

Finally, it's time to compile and run the pipeline code in the notebook. The notebook specifies the name of the run and the experiment (a group of runs), which is then displayed in the Kubeflow dashboard. By clicking on the notebook's run link , you can see the pipeline running in the Kubeflow Pipelines UI:

```
In [ ]:  experiment_name = 'fashion_mnist_kubeflow'
         run_name = pipeline_func.__name__ + ' run'

         arguments = {"data_path":DATA_PATH,
                      "model_file":MODEL_PATH,
                      "image_number": IMAGE_NUMBER}

         # Compile pipeline to generate compressed YAML definition of the pipeline.
         kfp.compiler.Compiler().compile(pipeline_func,
           '{}.zip'.format(experiment_name))

         # Submit pipeline directly from pipeline function
         run_result = client.create_run_from_pipeline_func(pipeline_func,
                                            experiment_name=experiment_name,
                                            run_name=run_name,
                                            arguments=arguments)
```

Figure 9.35 – Compiling and running the pipeline

Now that the pipeline has been created and set to run, we can look at its results. By clicking on the notebook's run link, we can get to the Kubeflow Pipelines dashboard. The pipeline's defined components will be displayed. The path of the data pipeline will be updated as they complete:

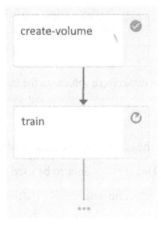

Figure 9.36 – Kubeflow Pipelines dashboard

To view the details of a component, we can click on it directly and navigate through a few tabs. To view the logs that were generated while running the component, go to the **Logs** tab:

```
1875/1875 [==============================] - 3s 2ms/step - loss: 0.2949 - accuracy: 0.8913
Epoch 6/10
1875/1875 [==============================] - 3s 2ms/step - loss: 0.2776 - accuracy: 0.8977
Epoch 7/10
1875/1875 [==============================] - 3s 2ms/step - loss: 0.2669 - accuracy: 0.9022
Epoch 8/10
1875/1875 [==============================] - 3s 2ms/step - loss: 0.2552 - accuracy: 0.9046
Epoch 9/10
1875/1875 [==============================] - 3s 2ms/step - loss: 0.2463 - accuracy: 0.9089
Epoch 10/10
1875/1875 [==============================] - 3s 2ms/step - loss: 0.2376 - accuracy: 0.9117
<keras.callbacks.History at 0x7f5f2c785110>
```

Figure 9.37 – Model training logs

The loss and accuracy metrics are displayed as the model trains. On the training data, the model achieves an accuracy of about 0.91 (or 91%), as shown in *Figure 9.37*.

Once the echo_result component has finished executing, you can inspect the component's logs to see what happened. It will show the class of the image being predicted, the model's confidence in its prediction, and the image's actual label:

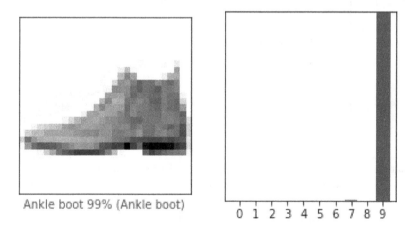

Ankle boot 99% (Ankle boot)

Figure 9.38 – Final prediction result from the pipeline

With that, we have made use of the Fashion MNIST dataset and TensorFlow's Basic classification to turn the example model into a Kubeflow pipeline.

Now, let's look at the best practices for running AI/ML workloads on Kubernetes.

Recommendations – running AL/ML workloads on Kubernetes

If you're a data scientist or ML engineer, you're probably thinking about how to deploy your ML models efficiently. You would essentially look for ways to scale models, distribute them across server clusters, and optimize model performance with a variety of techniques.

These are all tasks that Kubernetes is very good at. But Kubernetes was not designed to be an ML deployment platform. However, as more data scientists turn to Kubernetes to run their models, Kubernetes and ML are becoming popular stacks.

As a platform for training and deploying ML models, Kubernetes provides several key advantages. To understand those benefits, let's compare some of the major challenges and Kubernetes solution offerings:

Parameter	Challenges	How Kubernetes Offers a Competitive Edge
Need for reusable deployment resources	You'll have to deploy your models manually unless you set up an automated release pipeline. There is no easily reusable code or commands that you can re-invoke for each new deployment by default.	When you deploy a workload on Kubernetes, you can define it in a YAML or JSON file, or by creating a Helm chart. In either case, when it's time to redeploy, you can easily modify and reuse the deployment files. In this sense, Kubernetes provides an automated deployment solution by default, without requiring users to perform additional work or set up additional tools (more deployment automation could be achieved using a CI/CD pipeline that feeds into Kubernetes).
Scaling requirements	Model training frequently necessitates massive computing power. However, scaling up training by distributing the load across multiple servers can be difficult. This is because there is no simple way to scale a model running on one machine to another unless the machines are organized into a cluster.	To ensure that workloads have enough resources to run, Kubernetes automatically scales them up and distributes them across a cluster of servers.

Parameter	Challenges	How Kubernetes Offers a Competitive Edge
Management of infrastructure	If you distribute your models across multiple servers, you run the risk of training failing if one of those servers crashes and you don't have a way to automatically fail over to a different server.	When a server in the cluster begins to fail or stops responding, Kubernetes automatically redistributes workloads. This means that your training will not fail and will not have to be restarted every time a server fails.
Multi-tenancy requirements	If a data science team spends time and money setting up a cluster, the team will likely want to run multiple models on it at the same time. However, depending on how the cluster is managed, it may be difficult to isolate workloads from one another.	Kubernetes supports multi-tenancy, which allows users to isolate workloads from one another. Data scientists can create a single cluster and distribute it across multiple workloads or teams.
Gpu requirements	When you set up a cluster using virtual machines, it becomes more difficult to offload computation to GPUs.	Your models can directly access GPUs if you deploy them inside containers and properly provision your nodes. Even if you run models in virtual machines, GPU passthrough is supported by add-on frameworks such as KubeVirt, which allows you to orchestrate VMs with Kubernetes. As a result, regardless of how you run your models on Kubernetes, they can benefit from GPU offloading.

Table 9.2 – Kubernetes solution offerings

Kubernetes helps offset some of the most significant challenges that data scientists face when running models at scale in each of these ways. Now, let's look at some of the best practices for running AI/ML workloads.

Best practices for running AI/ML workloads

As ML progresses from research to practical applications, we must improve the maturity of its operational processes. To crack these challenges, we'll need to integrate DevOps and data engineering methods, as well as ones that are specific to ML.

MLOps blends ML, DevOps, and data engineering into a set of approaches. MLOps seeks to deploy and manage ML systems in production reliably and efficiently. The following are some of MLOps's main practices:

- **Data Pipelines**: ML models always require some type of data transformation, which is typically accomplished through scripts or even cells in a notebook, making them difficult to manage and run consistently. Creating a separate data pipeline offers numerous benefits in terms of code reuse, runtime visibility, management, and scalability.

- **Pipeline Versions**: Most ML models require two pipeline versions: one for training and one for serving. This is because data formats and the methods for accessing them are typically very different from each other, particularly for models that need to be served in real-time requests (as opposed to batch prediction runs). The ML pipeline should be a pure code artifact that is independent of any specific data. This means that it is possible to track its versions in source control and automate its deployment with a regular CI/CD pipeline. This would enable us to connect the code and data planes in a structured and automated manner.

- **Multiple Pipelines**: At this point, it's clear that there are two types of ML pipelines: training pipelines and serving pipelines. They have one thing in common: the data transformations that are performed must produce data in the same format, but their implementations can differ greatly. For example, the training pipeline typically runs over batch files containing all features, whereas the serving pipeline frequently runs online and receives only a portion of the features in the requests, retrieving the remainder from a database. It is critical, however, to ensure that these two pipelines are consistent, so code and data should be reused whenever possible.

- **Model and Data Versioning**: Consistent version tracking is essential for reproducibility. In a traditional software world, versioning code is sufficient because it defines all behavior. In ML, we must also keep track of model versions, as well as the data used to train them and some meta information such as training hyperparameters. It's also necessary to version the data and associate each trained model with the exact versions of code, data, and hyperparameters that we used.

- **Model Validation**: To determine whether a model is suitable for deployment, the right metrics to track and the threshold of acceptable values must be determined, usually empirically and frequently compared to previous models or benchmarks.

- **Data Validation**: A good data pipeline will typically begin by validating the input data. File format and size, column types, null or empty values, and invalid values are all common validations. All of these are required for ML training and prediction; otherwise, you may have a misbehaving model. Higher-level statistical properties of the input should also be validated by ML pipelines. For example, if the average or standard deviation of a feature varies significantly from one training dataset to the next, the trained model and its predictions will most likely be affected.

- **Monitoring**: Monitoring becomes important for ML systems because their performance is dependent not only on factors over which we have some control, such as infrastructure and our software, but also on data, over which we have much less control. In addition to standard metrics such as latency, traffic, errors, and saturation, we must also monitor model prediction performance. To detect problems that affect specific segments, we must monitor metrics across slices (rather than just globally), just as we do when validating the model.

To summarize, deploying ML in a production context entails more than just publishing the model as a prediction API. Rather, it entails establishing an ML pipeline capable of automating the retraining and deployment of new models. Setting up a CI/CD system allows you to test and release new pipeline implementations automatically. This system enables us to deal with quick data and business environment changes. MLOps, as a new area, is quickly gaining traction among data scientists, ML engineers, and AI enthusiasts.

Summary

To summarize, Kubeflow provides an easy-to-deploy, easy-to-use toolchain that will allow data scientists to integrate the various resources they will need to run models on Kubernetes, such as Jupyter Notebooks, Kubernetes deployment files, and ML libraries such as PyTorch and TensorFlow.

Another popular ML task that Kubeflow considerably simplifies is working with Jupyter Notebooks. You can build notebooks and share them with your team or teams using Kubeflow's built-in notebook services, which you can access via the UI. In this chapter, we learned how to set up an ML pipeline that will develop and deploy an example model using the Kubeflow ML platform. We also recognized that Kubeflow on MicroK8s is easy to set up and configure, as well as lightweight and capable of simulating real-world conditions while constructing, migrating, and deploying pipelines.

In the next chapter, you will learn how to deploy and run serverless applications using the Knative and OpenFaaS frameworks.

10

Going Serverless with Knative and OpenFaaS Frameworks

In the last chapter, we discussed Kubeflow, which provides an easy-to-deploy, simple-to-use toolchain for data scientists to integrate the various resources they will need to run models on Kubernetes, such as Jupyter notebooks, Kubernetes deployment files, and machine learning libraries such as PyTorch and TensorFlow.

By using Kubeflow's built-in Notebooks services, you can create notebooks and share them with your teams. We also went over how to set up a machine learning pipeline to develop and deploy an example model using the Kubeflow machine learning platform. Additionally, we established that Kubeflow on MicroK8s is simple to set up and configure, lightweight, and capable of simulating real-world conditions while building, migrating, and deploying pipelines.

In this chapter, we will look at the most popular open source serverless frameworks that extend Kubernetes with components for deploying, operating, and managing serverless, cloud-native apps. These frameworks enable you to create a service by encapsulating the code in a container image and delivering the required functionalities. Serverless frameworks automatically start and stop instances, so your code only runs when it's needed. Unless your code needs to accomplish something, resources aren't used.

Kubernetes' container orchestration capabilities (such as scheduling, load balancing, and health monitoring) make container proliferation much easier. However, this requires developers to perform or template several repetitive tasks, such as pulling application source code from repositories, building and provisioning a container image around the code, and configuring network connections outside of Kubernetes using various tools. Additionally, integrating Kubernetes-managed containers into an automated **continuous integration/continuous delivery** (**CI/CD**) pipeline necessitates the use of new tools and scripting.

With serverless frameworks automating the aforementioned activities from within Kubernetes, it eliminates complexity. A developer would be able to define the contents and configuration of a container in a single YAML manifest file, and serverless frameworks would take care of the rest, including building the container and conducting network programming to set up a route, should be Ingress, load balancing, and more.

Serverless computing is becoming the preferred cloud-native execution approach as it makes developing and running applications much easier and more cost-effective.

The serverless model of computing offers the following:

- Provisioning of resources on demand, scaling transparently based on demands, and scaling to zero when no more requests are made

- Offloading all infrastructure management responsibilities to the infrastructure provider, allowing developers to spend their time and effort on creation and innovation

- Allowing users to pay only for resources that are used, never for idle capacity

Kubernetes cannot run serverless apps on its own; we would need customized software that combines Kubernetes with a specific infrastructure provider's serverless platform. By abstracting away the code and handling network routing, event triggers, and autoscaling, the serverless frameworks would enable any container to run as a serverless workload on any Kubernetes cluster; it doesn't matter whether the container is built around a serverless function or other application code (for example, microservices).

Serverless computing, especially when deployed at the network's edge, is considered a key enabler for the building of increasingly complex **Internet of Things** (**IoT**) systems in the future. However, when installing new edge infrastructures for serverless workloads, additional attention must be paid to resource usage and network connectivity. Studies show that edge-oriented distributions, such as MicroK8s, perform better in the majority of tests, including cold start delay, serial execution performance, parallel execution with a single replica, and parallel execution using various autoscaling techniques.

We'll look at two of the most popular serverless frameworks included with MicroK8s in this chapter: Knative and OpenFaaS. Both serverless frameworks are Kubernetes-based platforms for building, deploying, and managing modern serverless workloads. In this chapter, we're going to cover the following main topics:

- Overview – Knative framework

- Enabling the Knative add-on

- Deploying and running a sample service on Knative

- Overview – OpenFaaS framework

- Enabling the OpenFaaS add-on

- Deploying and running a sample function on OpenFaaS

- Best practices for developing and deploying serverless applications

Overview of the Knative framework

Knative is a Kubernetes-based platform for deploying, managing, and scaling modern serverless workloads. Knative has the following three main components:

- **Build**: Provides streamlined source-to-container builds that are easy to utilize. By utilizing common constructs, you gain an advantage.

- **Serving**: Networking, autoscaling, and revision tracking are all handled by Knative. All you have to do now is concentrate on your core logic.

- **Eventing**: Handles the subscription, delivery, and management of events. By connecting containers to a data stream via declarative event connection and developer-friendly object architecture, you can create modern apps.

MicroK8s is the optimal solution to getting started with all of the components of Knative (Build, Serving, and Eventing) because it provides native support for Knative. We'll go through each component in detail in the next section.

Build components

The Knative Build component simplifies the process of building a container from source code. This procedure usually consists of the following steps:

1. Downloading source code from a code repository such as GitHub

2. Installing the underlying dependencies that the code requires to run, such as environment variables and software libraries

3. Container image creation

4. Placing container images in a registry accessible to the Kubernetes cluster

For its Build process, Knative makes use of Kubernetes **application programming interfaces (APIs)** and other technologies. The developer can use a single manifest (usually a YAML file) that describes all of the variables' location of the source code, required dependencies, and so on. Knative leverages the manifest to automate the container building and image creation process.

Serving components

Containers are deployed and run as scalable Knative services via the Serving component. The following are the key capabilities provided by the Serving component:

- **Configuration**: A service's state is defined and maintained by configuration. It also has version control. Each change to the configuration creates a new version of the service, which is saved alongside earlier versions.

- **Intelligent service routing**: Developers can use intelligent service routing to direct traffic to different versions of the service. Assume you've produced a new version of a service and want to test it out on a small group of users before moving everyone. Intelligent service routing allows you to send a portion of user requests to the new service and the rest to an older version. As you gain confidence in the new service, you can send more traffic to it.

- **Autoscaling**: Knative can scale services up to thousands of instances and down to zero instances, which is critical for serverless applications.

- **Istio** (`https://istio.io/`): This is an open source Kubernetes service mesh deployed along with Knative. It offers service request authentication, automatic traffic encryption for safe communication between services, and extensive metrics on microservices and serverless function operations for developers and administrators to use to improve infrastructure.

Knative Serving is defined by a set of objects known as Kubernetes **Custom Resource Definitions (CRDs)**. The following components define and govern the behavior of your serverless workload on the cluster:

- **Service**: Controls the entire life cycle of your workload for you. It ensures that your app has a route, a configuration, and a new revision for each service update by controlling the creation of additional objects. The service can be configured to always send traffic to the most recent revision or a pinned revision.

- **Route**: A network endpoint is mapped to one or more revisions. Traffic can be managed in a variety of ways, including fractional traffic and named routes.

- **Revision**: This is a snapshot of the code and configuration for each change made to the workload at a specific moment in time. Revisions are immutable objects that can be kept for as long as they are needed. Knative Serving Revisions can be scaled up and down automatically in response to incoming traffic.

- **Configuration**: This keeps your deployment in the desired state. It adheres to the Twelve-Factor App paradigm and provides a clear separation between code and configuration. A new revision is created when you change a configuration.

To summarize, the Serving component is responsible for deploying and running containers as scalable Knative services.

Eventing components

Knative's Eventing component allows various events to trigger container-based services and functions. There is no need to develop scripts or implement middleware because Knative queues handle the distribution of events to the respective containers. A messaging bus that distributes events to containers and channels, which are nothing but queues of events (from which developers can choose), is also handled by Knative. Developers can also establish feeds that connect an event to a specific action that their containers should execute.

Knative event sources make integration with third-party event providers easier for developers. The Eventing component will connect to the event producer and route the generated events automatically. It also provides tools for routing events from event producers to sinks, allowing developers to build applications that use an event-driven architecture.

Knative Eventing resources are loosely coupled and can be developed and deployed separately. Any producer can generate events and any event consumer can express interest in that event or group of events. Knative Eventing also takes care of sending and receiving events between event producers and sinks using standard HTTP POST requests. The following are the Eventing components:

- **Event sources**: These are the primary event producers in a Knative Eventing deployment. Events are routed to either a sink or a subscriber.

- **Brokers and Triggers**: These provide an event mesh model that allows event producers to deliver events to a Broker, which then distributes them uniformly to consumers via Triggers.

- **Channels and Subscriptions**: These work together to create an event pipe model that transforms and routes events between channels via Subscriptions. This model is suitable for event pipelines in which events from one system must be transformed before being routed to another process.

- **Event registry**: Knative Eventing defines an EventType object to help consumers discover the types of events available from Brokers. The registry is made up of various event types. The event types stored in the registry contain all of the information needed for a consumer to create a Trigger without using an out-of-band mechanism.

In the following figure, Knative components are represented. Serving and Eventing collaborate on tasks and applications to automate and manage them:

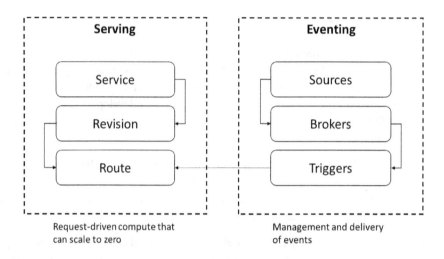

Figure 10.1 – Knative components

To recap, Knative provides components that allow the following:

- Serverless containers can be deployed quickly.
- Scaling pods down to zero as well as autoscaling based on demands.
- Multiple networking layers are supported for integration including Contour, Kourier, and Istio.
- Support for point-in-time snapshots of deployed code and configurations.
- Support for both HTTP and HTTPS networking protocols.

Now that we've covered the basics of Knative, we'll enable the add-on and deploy one of the samples in the next section.

Enabling the Knative add-on

Since Knative isn't available for ARM64 architecture, we will be using an Ubuntu virtual machine for this section. The instructions for setting up the MicroK8s cluster are the same as in *Chapter 5, Creating and Implementing Updates on Multi-Node Raspberry Pi Kubernetes Clusters*.

We'll enable the Knative add-on that adds Knative middleware to your cluster. Use the following command to enable the Knative add-on:

```
microk8s enable knative
```

When you enable this add-on, Istio and DNS will be also added to MicroK8s.

The following command execution output confirms that the Knative add-on is being enabled:

```
$ microk8s enable knative
Enabling Knative
Enabling Istio
Fetching istioctl version v1.10.3.
  % Total    % Received % Xferd  Average Speed   Time    Time     Time  Current
                                 Dload  Upload   Total   Spent    Left  Speed
100   668  100   668    0     0   2241      0 --:--:-- --:--:-- --:--:--  2241
100 21.3M  100 21.3M    0     0  15.3M      0  0:00:01  0:00:01 --:--:-- 31.1M
istio-1.10.3/
istio-1.10.3/manifests/
istio-1.10.3/manifests/charts/
istio-1.10.3/manifests/charts/istio-operator/
```

Figure 10.2 – Enabling the Knative add-on

It will take some time to finish activating the add-on. The following command execution output shows that Knative has been successfully enabled:

```
service/broker-ingress created
deployment.apps/mt-broker-controller created
Warning: autoscaling/v2beta2 HorizontalPodAutoscaler is deprecated in v1.23+, unavaila
 autoscaling/v2 HorizontalPodAutoscaler
horizontalpodautoscaler.autoscaling/broker-ingress-hpa created
horizontalpodautoscaler.autoscaling/broker-filter-hpa created


Visit https://knative.dev/docs/admin/ to customize which broker channel
implementation is used and to specify which configurations are used for which namespac

$
```

Figure 10.3 – Knative add-on activated

Before moving on to the next step, let's verify whether the add-on has been enabled and that all the required pods are running.

To see whether the add-on is activated or not, use the `kubectl get pods -n knative-serving` command. The following command execution output indicates that Knative Serving components are running:

```
$ kubectl get pods -n knative-serving
NAME                                      READY   STATUS    RESTARTS   AGE
autoscaler-74f697b6c6-bt4sl               1/1     Running   0          2m31s
controller-84f98b57b-jd2q7                1/1     Running   0          2m30s
activator-6b8d5bccb4-7sqrs                1/1     Running   0          2m31s
domain-mapping-69479cf66f-8742d           1/1     Running   0          2m30s
net-istio-controller-6b876996dc-9dkl6     1/1     Running   0          2m15s
net-istio-webhook-d45dbdcb6-15xs7         1/1     Running   0          2m15s
default-domain-fk62r                      1/1     Running   0          2m7s
domainmapping-webhook-bb67b5f65-8hx6p     1/1     Running   0          2m30s
webhook-5dcd765485-wf8ck                  1/1     Running   0          2m29s
$
```

Figure 10.4 – Knative Serving component pods are running

Before we move on to the next step, let's make sure that all of the Knative Eventing components are up and running using the following command:

```
kubectl get pods -n knative-eventing
```

The following command execution output indicates that Knative Eventing components are also running:

```
$ kubectl get pods -n knative-eventing
NAME                                      READY   STATUS    RESTARTS   AGE
eventing-controller-7bfd95cc79-ztx4z      1/1     Running   0          2m39s
eventing-webhook-c7998d8f9-gcj55          1/1     Running   0          2m38s
imc-dispatcher-dd5bbb4d7-vrqj7            1/1     Running   0          2m13s
imc-controller-6f74957b95-g5x9s           1/1     Running   0          2m18s
mt-broker-controller-658f88d698-j7wtd     1/1     Running   0          89s
mt-broker-filter-5fd68bd989-d7dcp         1/1     Running   0          91s
mt-broker-ingress-5bd6749895-kdlmk        1/1     Running   0          90s
$
```

Figure 10.5 – Knative Eventing components are running

We now have all of the components of Knative up and running.

We will proceed to the next step of installing the Knative **command-line interface (CLI)** tool kn. Without having to create or edit YAML files manually, kn provides a quick and easy interface for building Knative resources, such as services and event sources. It also makes it easier to complete tasks such as autoscaling and traffic splitting that might otherwise be difficult.

The kn binary can be downloaded from the release page (`https://github.com/knative/client/releases`) and copied to `/usr/local/bin` using the following command:

```
sudo curl -o /usr/local/bin/kn -sL https://github.com/knative/
client/releases/download/knative-v1.3.1kn-linux-amd64
```

The following command execution output confirms that the kn CLI has been downloaded successfully and is available at `/usr/local/bin`:

```
$ sudo curl -o /usr/local/bin/kn -sL https://github.com/knative/client/releases/download/knative-v1.3.1
kn-linux-amd64
$ sudo chmod +x /usr/local/bin/kn
$
```

Figure 10.6 – Installing the Knative CLI

Before moving on to the next step, let's verify whether the kn CLI is working by running the following `kn version` command:

```
kn version
```

The following output confirms that the kn CLI is operational, and its version and build date are displayed:

```
$ kn version
Version:      v1.3.1
Build Date:   2022-03-11 18:43:10
Git Revision: a591c0c0
Supported APIs:
* Serving
  - serving.knative.dev/v1 (knative-serving v1.3.0)
* Eventing
  - sources.knative.dev/v1 (knative-eventing v1.3.0)
  - eventing.knative.dev/v1 (knative-eventing v1.3.0)
$
```

Figure 10.7 – Verifying whether the kn CLI is operational

For the kn CLI to access Kubernetes configuration, copy the MicroK8s configuration file to `$HOME/.kube/config` as follows:

```
$ microk8s config > $HOME/.kube/config
```

Figure 10.8 – Copy MicroK8s configuration file to $HOME folder

All of the Knative components, as well as the Knative CLI kn setup, are now up and running. We'll now move on to the following step: deploying and running the sample service.

Deploying and running a sample service on Knative

In this section, we will deploy the `Hello world` sample service from the Knative samples repo. The sample service prints `Hello $TARGET!` after reading the `TARGET` environment variable. If `TARGET` is not given, the default value is "`World`".

Now in the following steps, we'll deploy the service by specifying the image location and the `TARGET` environment variable. We are going to create a Knative service (Serving component), which is a time-based representation of a single serverless container environment (such as a microservice). It includes both the network address for accessing the service and the application code and settings required to run the service.

A Knative service lifespan is controlled by the `serving.knative.dev` CRD. To create the Knative service, we'll use the `kn` CLI as follows:

```
kn service create kn-serverless --image gcr.io/knative-samples/
helloworld-go --env TARGET=upnxtblog.com
```

The following command execution output indicates that the service creation is successful and the service can be accessed at the URL `http//kn-serverless.default.example.com`:

```
$ kn service create kn-serverless --image gcr.io/knative-samples/helloworld-go --env TARGET=upnxtblog.com
Creating service 'kn-serverless' in namespace 'default':

  0.083s The Route is still working to reflect the latest desired specification.
  0.105s Configuration "kn-serverless" is waiting for a Revision to become ready.
  0.118s ...
 45.781s ...
 45.873s Ingress has not yet been reconciled.
 45.952s Waiting for Envoys to receive Endpoints data.
 46.260s Waiting for load balancer to be ready
 46.519s Ready to serve.

Service 'kn-serverless' created to latest revision 'kn-serverless-00001' is available at URL:
http://kn-serverless.default.example.com
$ 
```

Figure 10.9 – Creating a new Knative service

Congrats! We have successfully created a new Knative service and deployed it.

The following is a recap of the Serving components:

- **Service**: Manages the whole life cycle of your workload
- **Route**: Takes care of mapping the network endpoint to one or more revisions
- **Configuration**: Maintains the desired state for the deployment
- **Revision**: A point-in-time snapshot of the code and configuration of the workload

The Serving components involved in the definition and control of how serverless workloads behave on the cluster are depicted in the following diagram:

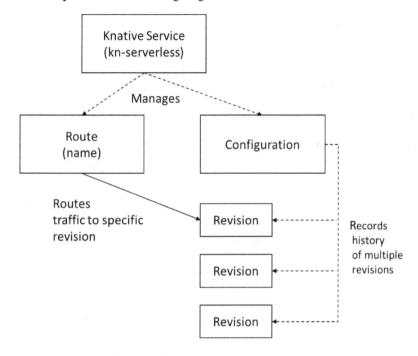

Figure 10.10 – Knative Serving components

We can now invoke the service that we previously created using the `curl` command as follows:

```
curl http://$SERVICE_IP:$INGRESS_PORT/ -H 'Host: kn-serverless.
default.example.com'
```

$SERVICE_IP and $INGRESS_PORT point to the Knative service and the Ingress port that is exposed. The output of the following command confirms that the Knative service has been invoked and output has been displayed:

```
$ curl http://$SERVICE_IP:$INGRESS_PORT/ -H 'Host: kn-serverless.default.example.com'
Hello upnxtblog.com!
$
```

Figure 10.11 – Invoking the Knative service

To observe how a pod is created to service the requests, run the `watch kubectl get pods` command in a new Terminal tab. If there are no inbound requests for 60 seconds, Knative will automatically scale this pod down to zero as shown in the following command execution output:

```
Every 2.0s: kubectl get pods
0:00 Fri Apr  1 06:19:45 2022
NAME                                                READY   STATUS        RESTARTS
kn-serverless-00001-deployment-57bf78bd47-tz4gc     2/2     Terminating   0
s
```

Figure 10.12 – Pods are terminated if there are no inbound requests for 60 seconds

You may also issue the preceding `curl` command after the pods have scaled down to zero to watch the pod spin up and serve the request zero as shown in the following command execution output:

```
$ kubectl get pods
NAME                                                READY   STATUS    RESTARTS   AGE
kn-serverless-00001-deployment-57bf78bd47-tz4gc     2/2     Running   0          109s
$
```

Figure 10.13 – Pods are spun up to serve the requests

Knative, in a nutshell, is a Kubernetes-powered platform for developing, deploying, and managing modern serverless workloads. We also discovered that MicroK8s has native Knative support and is the best way to get started with all of Knative's components (Build, Serving, and Eventing).

We have deployed a sample application and used its endpoints to call it from the command line. We will now look at the next choice, OpenFaaS, in the next section to run the sample application, and analyze the features it offers.

Overview of the OpenFaaS framework

OpenFaaS (**FaaS** standing for **functions as a service**) is a framework for creating serverless functions using the Docker and Kubernetes container technologies. Any process can be packaged as a function, allowing the consumption of a variety of web events without having to write boilerplate code over and over. It's an open source initiative that's gaining a lot of traction in the community.

Some of the key advantages of the OpenFaaS framework are the following:

- Running functions on any infrastructure without concern of lock-in with an open source functions framework.

- Creating functions in any programming language and packaging them in Docker/OCI containers.

- Built-in UI, robust CLI, and one-click installation make it simple to use.

- Scale as you go – handle traffic spikes and scale down when not in use.

- A community edition and a pro edition are available along with production support.

Now that we've covered the concepts of OpenFaaS, we'll enable the add-on and deploy one of the samples in the next section.

Enabling the OpenFaaS add-on

Since OpenFaaS isn't available for ARM64 architecture, we will be using an Ubuntu virtual machine for this section. The instructions for setting up the MicroK8s cluster are the same as in *Chapter 5, Creating and Implementing Updates on Multi-Node Raspberry Pi Kubernetes Clusters.*

Before enabling the OpenFaaS add-on, enable the DNS and Registry add-ons using the following command:

```
microk8s enable dns
```

The DNS is used to provide address resolution services to Kubernetes so that services can communicate with each other. The following command execution output confirms that the DNS add-on is enabled:

```
$ microk8s enable dns
Enabling DNS
Applying manifest
serviceaccount/coredns created
configmap/coredns created
deployment.apps/coredns created
service/kube-dns created
clusterrole.rbac.authorization.k8s.io/coredns created
clusterrolebinding.rbac.authorization.k8s.io/coredns created
Restarting kubelet
DNS is enabled
$ 
```

Figure 10.14 – Enabling the DNS add-on

Now that the DNS add-on is enabled, we will move on to the next step of enabling the Registry add-on using the following command:

```
microk8s enable registry
```

The Registry add-on creates a private registry in Docker and exposes it at `localhost:32000`. As part of this add-on, the storage add-on will also be enabled as follows:

```
$ microk8s enable registry
The registry will be created with the default size of 20Gi.
You can use the "size" argument while enabling the registry, eg microk8s.enabl
Enabling default storage class
deployment.apps/hostpath-provisioner created
storageclass.storage.k8s.io/microk8s-hostpath created
serviceaccount/microk8s-hostpath created
clusterrole.rbac.authorization.k8s.io/microk8s-hostpath created
clusterrolebinding.rbac.authorization.k8s.io/microk8s-hostpath created
Storage will be available soon
Applying registry manifest
namespace/container-registry created
persistentvolumeclaim/registry-claim created
deployment.apps/registry created
service/registry created
configmap/local-registry-hosting configured
The registry is enabled
$
```

Figure 10.15 – Enabling the Registry add-on

We can move on to the next step of enabling the OpenFaaS add-on now that we've enabled the DNS and Registry add-ons.

Use the following command to enable the OpenFaaS add-on:

```
microk8s enable openfaas
```

The following command execution output confirms that the OpenFaaS add-on is being enabled:

```
$ microk8s enable openfaas
Addon dns is already enabled.
Enabling Helm 3
Fetching helm version v3.5.0.
  % Total    % Received % Xferd  Average Speed   Time    Time     Time  Current
                                 Dload  Upload   Total   Spent    Left  Speed
100 11.7M  100 11.7M    0     0  12.8M      0 --:--:-- --:--:-- --:--:-- 12.8M
Helm 3 is enabled

Enabling OpenFaaS
Operator: false
Basic Auth enabled: true
WARNING: Kubernetes configuration file is group-readable. This is insecure. Loca
3052/credentials/client.config
```

Figure 10.16 – Enabling the OpenFaaS add-on

It will take some time to finish activating the add-on. The following command execution output shows that the OpenFaaS add-on has been successfully enabled:

```
REVISION: 1
TEST SUITE: None
NOTES:
To verify that openfaas has started, run:

  kubectl -n openfaas get deployments -l
To retrieve the admin password, run:

  echo $(kubectl -n openfaas get secret ba
ecode)
OpenFaaS has been installed
$ 
```

Figure 10.17 – The OpenFaaS add-on is enabled

As you can see, deployment scripts generate a username (admin) and password combination during the installation. Save the credentials so we can use them in the following steps.

Before moving on to the next step, let's verify whether the add-on has been enabled and that all the required pods are running.

To see whether the add-on is activated, use the kubectl get pods command. The following command execution output indicates that OpenFaaS components are running:

```
$ kubectl -n openfaas get deployments -l "release=openfaas, app=openfaas"
NAME                READY   UP-TO-DATE   AVAILABLE   AGE
prometheus          1/1     1            1           92s
nats                1/1     1            1           92s
queue-worker        1/1     1            1           92s
basic-auth-plugin   1/1     1            1           92s
alertmanager        1/1     1            1           92s
gateway             1/1     1            1           92s
$ 
```

Figure 10.18 – OpenFaaS pods are running

We now have all of the following components of OpenFaaS up and running:

1. nats provides asynchronous execution and queuing.

2. prometheus provides metrics and enables auto-scaling through alertmanager.

3. gateway provides an external route into the functions and also scales functions according to demand.

4. queue-worker is in charge of handling asynchronous requests.

We will proceed to the next step of installing the OpenFaaS CLI tool. The CLI can be used to create and deploy OpenFaaS functions. From a set of supported language templates, you can create OpenFaaS functions (such as Node.js, Python, C#, and Ruby). Please see the list of supported languages at `https://github.com/openfaas/templates` for further information.

You can use the `curl` command to install the CLI after acquiring the binaries from the releases page as follows:

```
curl -sSL -- insecure https://cli.openfaas.com | sudo -E sh
```

> **Note**
>
> Here we are using the `-insecure` flag to avoid any certificate download issues.

The following command execution output confirms that the CLI installation is successful. The `faas-cli` command and the `faas` alias are available post-installation as follows:

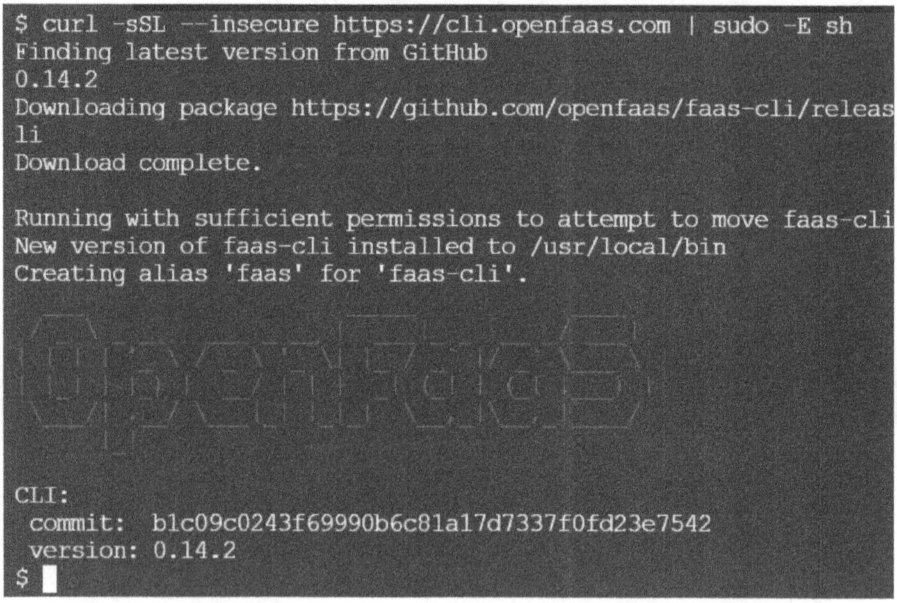

```
$ curl -sSL --insecure https://cli.openfaas.com | sudo -E sh
Finding latest version from GitHub
0.14.2
Downloading package https://github.com/openfaas/faas-cli/releas
li
Download complete.

Running with sufficient permissions to attempt to move faas-cli
New version of faas-cli installed to /usr/local/bin
Creating alias 'faas' for 'faas-cli'.




CLI:
  commit:  b1c09c0243f69990b6c81a17d7337f0fd23e7542
  version: 0.14.2
$
```

Figure 10.19 – Installing the OpenFaaS CLI

Now that we've installed `faas-cli`, we can use the `faas-cli` command to start creating and deploying functions in the next section

Deploying and running a sample function on OpenFaaS

This section will cover the creation, build, and deployment of a new FaaS Python function. We'll also use OpenFaaS CLI commands to test the deployed function. The OpenFaaS CLI has a template engine that can be used to set up new functions in any programming language. To create a new function, use the following command:

```
faas-cli new -lang <language template> --prefix localhost:32000
<function name>
```

Here -prefix localhost:32000 refers to the local MicroK8s registry that we have enabled in the preceding steps.

This command works by reading a list of templates from the ./template directory in your current working folder.

You can also use the faas-cli template pull command to pull the templates from the official OpenFaaS language templates from GitHub.

To check the list of languages that are supported, use the faas-cli new -list command.

The following command execution indicates that the new openfaas-serverless Python function has been created:

Figure 10.20 – Creating a new function using the CLI

A stack file and a new folder with the function name are generated in the current working folder as follows:

Figure 10.21 – A stack file and a new folder with the function name

Now that we've created a new function, we'll need to build it so that a container image can be created and used in the following steps.

Use the following command to build the new function:

```
faas-cli build -f ./openfaas-serverless.yml
```

The `faas-cli build` command creates a Docker image on your local MicroK8s registry, which could be used locally or could be uploaded to a remote container registry (in case of a multi-node cluster setup). Each change to your function necessitates issuing a new `faas-cli build` command.

The following command execution indicates that the new `openfaas-serverless` Python function has been built successfully:

```
$ faas-cli build -f ./openfaas-serverless.yml
[0] > Building openfaas-serverless.
Clearing temporary build folder: ./build/openfaas-
Preparing: ./openfaas-serverless/ build/openfaas-s
Building: localhost:32000/openfaas-serverless:late
Sending build context to Docker daemon  8.192kB
Step 1/31 : FROM --platform=${TARGETPLATFORM:-linu
chdog
 ---> 6f97aa96da81
Step 2/31 : FROM --platform=${TARGETPLATFORM:-linu
2.7-alpine: Pulling from library/python
aad63a933944: Pulling fs layer
259d822268fb: Pulling fs layer
10ba96d218d3: Pulling fs layer
44ba9f6a4209: Pulling fs layer
44ba9f6a4209: Waiting
```

Figure 10.22 – Building a new OpenFaaS function

It may take some time to complete the build process, but once completed, you should see the following output:

```
 ---> 3202cbddcb56
Step 31/31 : CMD ["fwatchdog"]
 ---> Running in bd9c86d0aaf9
Removing intermediate container bd9c86d0aaf9
 ---> b3af5c4a4e49
Successfully built b3af5c4a4e49
Successfully tagged localhost:32000/openfaas-serve
Image: localhost:32000/openfaas-serverless:latest
[0] < Building openfaas-serverless done in 35.92s.
[0] Worker done.

Total build time: 35.92s
$
```

Figure 10.23 – Successful OpenFaaS function build

We can move on to the next step of pushing the Docker image to the registry now that the images have been built.

Use the following command to push Docker images to our local registry:

```
faas-cli push -f ./openfaas-serverless.yml
```

The following command execution output indicates that the openfaas-serverless function has been pushed to the registry successfully:

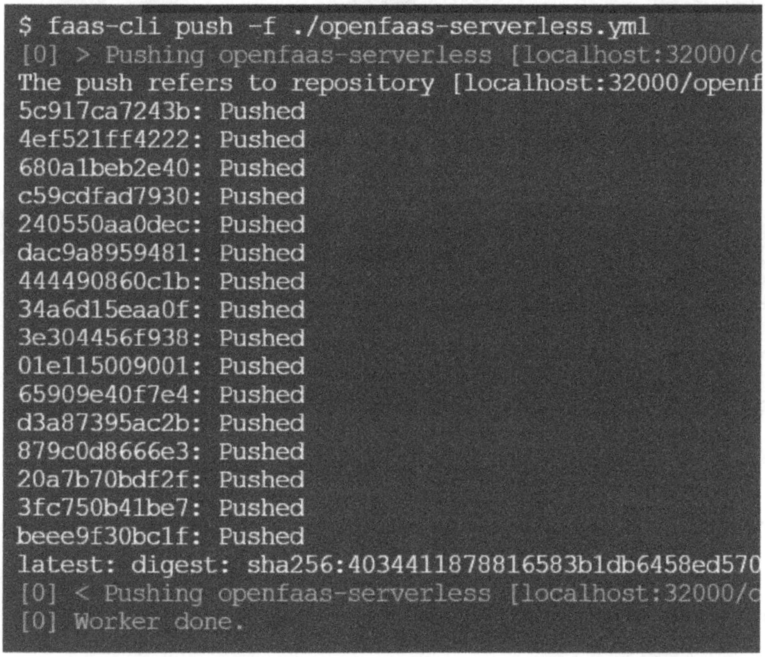

```
$ faas-cli push -f ./openfaas-serverless.yml
[0] > Pushing openfaas-serverless [localhost:32000/c
The push refers to repository [localhost:32000/openf
5c917ca7243b: Pushed
4ef521ff4222: Pushed
680a1beb2e40: Pushed
c59cdfad7930: Pushed
240550aa0dec: Pushed
dac9a8959481: Pushed
444490860c1b: Pushed
34a6d15eaa0f: Pushed
3e304456f938: Pushed
01e115009001: Pushed
65909e40f7e4: Pushed
d3a87395ac2b: Pushed
879c0d8666e3: Pushed
20a7b70bdf2f: Pushed
3fc750b41be7: Pushed
beee9f30bc1f: Pushed
latest: digest: sha256:4034411878816583b1db6458ed570
[0] < Pushing openfaas-serverless [localhost:32000/c
[0] Worker done.
```

Figure 10.24 – OpenFaaS function pushed to the local registry

Let's set an OPENFAAS_URL environment variable and also retrieve the necessary admin credentials before moving on to the next step of deploying the function.

An OPENFAAS_URL environment variable defines the default gateway URL that the CLI uses to contact the OpenFaaS server as follows:

```
$ export OPENFAAS_URL=http://127.0.0.1:31112
$
```

Figure 10.25 – Set the OPENFAAS_URL environment variable

To retrieve the admin credentials, use the following command that was printed during the installation:

```
echo $(kubectl -n openfaas get secret basic-auth -o
jsonpath="{.data.basic-auth-password}" | base64 --decode)
```

The following command execution output indicates that the command was successfully executed and that the password was retrieved:

```
$ echo $(kubectl -n openfaas get secret basic-auth -o jsonpath="{.data.basic-auth-password}" | base64 --d
ecode)
V5NjmSMhIb2q
```

Figure 10.26 – Retrieving admin credentials

Let's log in to the OpenFaaS server with the admin credentials so we can deploy the function. The following command execution output indicates that the login was successful and the credentials were saved to the local store:

```
$ faas-cli login admin
Must provide a non-empty password via --password or --password-stdin
$ faas-cli login admin --password V5NjmSMhIb2q
WARNING! Using --password is insecure, consider using: cat ~/faas_pass.
sword-stdin
Calling the OpenFaaS server to validate the credentials...
credentials saved for admin http://127.0.0.1:31112
$
```

Figure 10.27 – Using the retrieved password to log in to the OpenFaaS server

We can proceed to the next step of deploying the function to the OpenFaaS server now that the credentials have been saved.

Use the following command to deploy the function:

```
faas-cli deploy -f ./openfaas-serverless.yml
```

The following command execution output indicates that the deployment is successful and we now have the URL for accessing the function:

```
$ faas-cli deploy -f ./openfaas-serverless.yml
Deploying: openfaas-serverless.

Deployed. 202 Accepted.
URL: http://127.0.0.1:31112/function/openfaas-serverless

$
```

Figure 10.28 – Successful function deployment

Alternatively, you can use the `faas-cli up` command to build, push, and deploy the function in a single command.

Congrats! We have successfully created a new function and deployed it.

To call the function, we'll utilize the CLI's `invoke` function as follows:

```
faas-cli invoke -f openfaas-serverless.yml openfaas-serverless
```

By default, the function accepts an input parameter and outputs the input parameter value. To change the logic, the stack file and the handler file need to be modified, and then the function needs to be redeployed:

```
$ echo "Hello Upnxtblog.com" | faas-cli invoke -f openfaas-serverless.yml openfaas-serverless
Hello Upnxtblog.com

$
```

Figure 10.29 – Invoking the function

You could also use the OpenFaaS UI to invoke the deployed functions as follows:

Figure 10.30 – OpenFaaS UI

To summarize, in simple terms, OpenFaaS provides the following:

- A simple approach to package any code or binary, as well as a diverse ecosystem of language templates
- Built-in autoscaling and a function repository for collaboration and sharing metrics
- A Kubernetes-native experience and a devoted community

The best practices for developing and deploying serverless apps will be discussed in the following section.

Best practices for developing and deploying serverless applications

We must adhere to best practices in order to safeguard our resources, applications, and infrastructure service provider accounts. Here are some guiding principles that need to be considered.

Serverless function = specific function

A serverless function must accomplish a certain task. A serverless function should execute a logical function, similar to how a function or method in any code should accomplish one thing.

Using microservices

Microservices enable us to link together the data storage and functions in a manageable manner. The microservice will be bound by a contract that specifies what it is allowed and prohibited to do. A payment microservice, for example, can be used to create, update, and delete user payments. Outside of the user account data storage, this microservice should never modify any data. It will also have its own API. Other microservices can now interact with user account serverless functions in a consistent manner without modifying any of the user account data stores.

Using appropriate stacks for various resources

When deploying resources, serverless frameworks allow us to employ several language stacks, and each framework configuration deploys the appropriate stack. One stack per resource type should be the goal. Our user payment microservice, for example, might have a database stack (to store account metadata in MongoDB), an **identity provider** (**IdP**) stack (to set up and maintain user sessions with an OAuth2 provider), a function stack (to deploy functions that provide the user payment microservice API), and an object store stack (to capture user account profile pictures in S3). This enables us to edit one resource type without affecting another. If you make a mistake in the deployment of a function's stack, for example, your other stacks are unaffected.

Applying the principle of least privilege

The minimal set of IAM permissions should be applied to all of your resources. A serverless function that reads a MongoDB table, for example, should only contain the read action for that MongoDB table. When defining privileges, you should avoid using an asterisk (*) whenever feasible. A hacker can read and delete all database data if your function is ever compromised and it employs asterisks to make all MongoDB accessible and every operation permissible.

Performing load testing

Load testing your serverless functions would help you identify how much memory to allocate and what timeout value to use. In a serverless environment, there could be complicated apps, and you may not be aware of dependencies inside applications that prevent them from performing a function on heavy loads. Load testing allows you to identify possible problems that are critical to running a high-availability application.

Using a CI/CD pipeline

It's fine to deploy using the CLI when you're starting to build an application. Ideally, you should use a CI/CD pipeline to deploy your code before releasing it to production. Before enabling a pull request to merge, the CI section of the pipeline allows you to perform linting checks, unit tests, and a variety of additional automated checks. When a PR is merged or a branch is updated, the CD section of the pipeline allows you to deploy your serverless application automatically. Using a CI/CD pipeline eliminates human error and ensures that your process is repeatable.

Constant monitoring is required

We should monitor our serverless resources using services such as Knative monitoring and Prometheus. There may be many resources and they may be used so frequently that manually checking them for faults would be difficult. Health, longer executions, delays, and errors can all be reported by monitoring services. Having a service that alerts us (such as Alert Manager) when our serverless application and resources are experiencing problems allows us to locate and resolve issues more quickly.

Auditing in addition to monitoring

We want to audit in addition to monitoring. When anything stops working or has problems, monitoring alerts you. When our resources stray from a known configuration or are wrongly configured, auditing alerts us. We may develop rules that audit our resources and their configurations using services such as Knative Config or an OpenFaaS stack file.

Auditing software dependencies

We'd like to audit our software dependencies as well. Just because we don't have a server anymore doesn't mean we're immune to "patching." We want to make sure that any software dependencies we specify are current and do not include any known vulnerabilities. We can utilize automated tools to keep track of which software packages need to be updated.

Summary

In this chapter, we examined two of the most popular serverless frameworks included with MicroK8s, Knative and OpenFaaS, both of which are Kubernetes-based platforms for developing, deploying, and managing modern serverless workloads. We've deployed a few of the samples and used their endpoints to invoke them via the CLI. We also looked at how serverless frameworks scale down pods to zero when there are no requests and spin up new pods when there are more requests.

We realized that the ease of deployment of MicroK8s appears to be related to the ease with which serverless frameworks can be implemented. We've also discussed some guiding principles to keep in mind when developing and deploying serverless applications. However, deploying serverless resources is pretty simple. We also understood that in order to protect our resources, apps, and infrastructure service provider accounts, we needed to adhere to best practices.

In the next chapter, we'll look at how to use OpenEBS to implement storage replication that synchronizes data across several nodes.

Part 4:
Deploying and Managing Applications on MicroK8s

This part focuses on the deployment and management aspects of typical IoT/Edge computing applications, such as setting up storage replication for your stateful applications, implementing a service mesh for cross-cutting concerns and a high availability cluster to withstand a component failure and continue to serve workloads without interruption, configuring containers with workload isolation, and running secured containers in isolation from a host system.

This part of the book comprises the following chapters:

- *Chapter 11, Managing Storage Replication with OpenEBS*
- *Chapter 12, Implementing Service Mesh for Cross-Cutting Concerns*
- *Chapter 13, Resisting Component Failure Using HA Cluster*
- *Chapter 14, Hardware Virtualization for Securing Containers*
- *Chapter 15, Implementing Strict Confinement for Isolated Containers*
- *Chapter 16, Diving into the Future*

11
Managing Storage Replication with OpenEBS

In the previous chapter, we looked at two serverless frameworks that are available with MicroK8s, both of which are Kubernetes-based platforms for designing, deploying, and managing modern serverless workloads. We also noticed that the ease with which serverless frameworks can be implemented appears to be tied to the ease with which MicroK8s can be deployed. Some guiding principles to remember when creating and deploying serverless apps were also highlighted. We also realized that we needed to follow best practices to safeguard our resources, apps, and infrastructure service provider accounts.

In this chapter, we will look into the next use case for supporting cloud-native storage solutions, such as OpenEBS, to provide persistent storage for our container applications. Cloud-native storage solutions enable comprehensive storage mechanisms. These solutions mimic the properties of cloud environments, such as scalability, reliability, container architecture, and high availability. These features make it simple to interface with the container management platform and provide persistent storage for container-based applications.

First, we will look at the Kubernetes storage basics before diving into OpenEBS concepts. Containers are ephemeral, which means they are established for a specific reason and then shut down after that task is completed. Containers do not maintain state data on their own, and a new container instance has no memory/state of prior ones. Although a container provides storage, it is only ephemeral storage, so it is wiped when the container is turned off. Developers will need to manage persistent storage as part of containerized applications as they adopt containers for new use cases. A developer, for example, may want to operate a database in a container and store the data in a volume that survives the container's shutdown process.

Kubernetes provides a variety of management options for clusters of containers. The ability to manage persistent storage is one of these capabilities. Administrators can use Kubernetes persistent storage to keep track of both persistent and non-persistent data in a Kubernetes cluster. Multiple applications that operate on the cluster can then utilize storage resources dynamically.

To help manage persistent storage, Kubernetes supports two primary mechanisms:

- A **PersistentVolume** (**PV**) is a storage element that can be created manually or dynamically, depending on the storage class. It has a life cycle that is unaffected by the life cycle of Kubernetes pods. A pod can mount a PV, but the PV remains after the pod has shut down, and its data can still be accessed. Each PV can have its own set of parameters, such as disc type, storage tier, and performance.

- A **PersistentVolumeClaim** (**PVC**) is a storage request that's made by a Kubernetes user. Based on the custom parameters, any application operating on a container can request storage and define the size and other properties of the storage it requires (for example, the specific type of storage, such as SSD storage). Based on the available storage resources, the Kubernetes cluster can provision a PV.

`StorageClass` is a Kubernetes API object for configuring storage parameters. It's a way of configuring a dynamic setup that generates new volumes based on demand. `StorageClass` defines the name of the volume plugin, as well as any external providers and a **Container Storage Interface** (**CSI**) driver, which allows containers to communicate with storage devices. CSI is a standard that allows containerized workloads to access any block and file storage systems.

`StorageClass` can be defined and PVs assigned by Kubernetes administrators. Each `StorageClass` denotes a different form of storage, such as fast SSD storage versus traditional magnetic drives or remote cloud storage. This enables a Kubernetes cluster to supply different types of storage based on the workload's changing requirements.

Dynamic volume provisioning is a feature of Kubernetes that allows storage volumes to be created on-demand. Administrators no longer need to manually build new storage volumes in their cloud or storage provider, then create PV objects to make them available in the cluster. When users request a specific storage type, the entire process is automated and provisioned. `StorageClass` objects are defined by the cluster administrator as needed. A volume plugin such as OpenEBS, also known as a provisioner, is referenced by each `StorageClass`. When a storage volume is automatically provisioned, the volume plugin provides a set of parameters and passes them to the provisioner.

The administrator can define many `StorageClass`, each of which can represent a distinct type of storage or the same storage with different specifications. This allows users to choose from a variety of storage solutions without having to worry about the implementation details.

Container Attached Storage (**CAS**) is quickly gaining traction as a viable option for managing stateful workloads and is becoming the favored method for executing durable, fault-tolerant stateful applications. CAS was brought to the Kubernetes platform via the OpenEBS project. It can be readily deployed in on-premises clusters, managed clusters in the public cloud, and even isolated air-gapped clusters. MicroK8s offers in-built support for OpenEBS via an add-on, making it the best solution for running Kubernetes clusters in air-gapped Edge/IoT scenarios.

In this chapter, we're going to cover the following main topics:

- Overview of OpenEBS
- Configuring and implementing a PostgreSQL stateful workload
- Kubernetes storage best practices

Overview of OpenEBS

In Kubernetes, storage is often integrated as an OS kernel module with individual nodes. Even the PVs are monolithic and legacy resources since they are strongly tied to the underlying components. CAS allows Kubernetes users to treat storage entities as microservices. CAS is made up of two parts: the control plane and the data plane. The control plane is implemented as a set of **Custom Resource Definitions (CRDs)** that deal with low-level storage entities. The data plane runs as a collection of pods close to the workload. It is in charge of the I/O transactions, which translate into read and write operations.

The clean separation of the control plane and data plane provides the same benefits as running microservices on Kubernetes. This architecture decouples persistence from the underlying storage entities, allowing workloads to be more portable. It also adds scale-out capabilities to storage, allowing administrators and operators to dynamically expand volumes in response to the workload. Finally, CAS ensures that the data (PV) and compute (pod) are always co-located in a hyper-converged mode to maximize throughput and fault tolerance.

Data is copied across many nodes using the synchronous replication feature of OpenEBS. The failure of a node would only affect the volume replicas on that node. The data on other nodes would remain available at the same performance levels, allowing applications to be more resilient to failures:

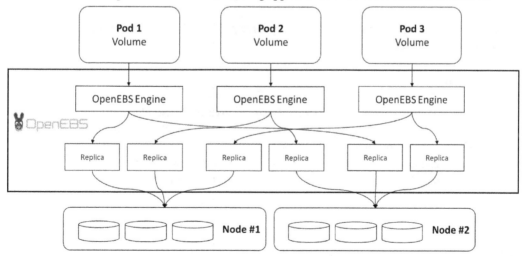

Figure 11.1 – Synchronous replication

Creating instantaneous snapshots are also possible with the OpenEBS CAS architecture. These can be made and managed with the regular `kubectl` command. This extensive integration with Kubernetes allows for job portability and easier data backup and migration.

The following diagram shows the typical components of OpenEBS:

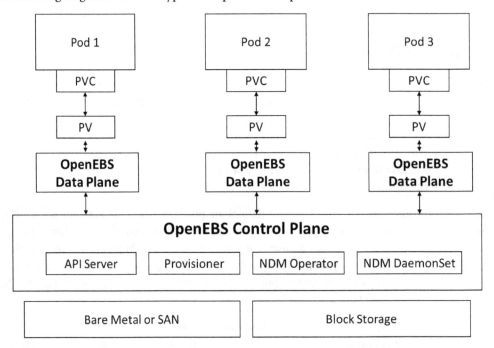

Figure 11.2 – OpenEBS control plane and data plane

OpenEBS is a well-designed system built on CAS concepts. We'll look at the architecture in more detail in the following sections.

Control plane

The control plane, disk manager, and data plane are assigned to each storage volume that's been installed. The control plane is closer to the storage infrastructure; it keeps track of the storage volumes that are joined to each cluster node through SAN or block storage. Provisioning volumes, initiating snapshots, creating clones, creating storage policies, enforcing storage policies, and exporting volume metrics to other systems such as Prometheus are all handled directly by the control plane.

An OpenEBS storage administrator interacts with the control plane to manage cluster-wide storage activities. Through an API server, the OpenEBS control plane is accessible to the outside world. A pod exposes the REST API for controlling resources such as volumes and policies. The declaration is initially submitted as a YAML file to the API server, which then starts the workflow. The API server communicates with the Kubernetes master's API server to schedule volume pods in the data plane.

Dynamic provisioning is implemented via the control plane's provisioner component using the standard Kubernetes external storage plugin. When an application builds a PVC from an existing storage class, the OpenEBS provisioner constructs a PV from the primitives in the storage class and binds it to the PVC.

The OpenEBS control plane relies heavily on the **Node Device Manager** (NDM). Each node in the Kubernetes cluster runs an NDM DaemonSet that is responsible for identifying new block storage devices and reporting them to the NDM operator to be registered as a block device resource if they meet the filter. NDM serves as a link between the control plane and the physical discs that each node is connected to. It keeps track of all registered block storage devices in the `etcd` database, which serves as the cluster's single source of truth.

Now that we've seen what the control plane it's, let's learn more about the data plane.

Data plane

The data plane is close to the workload, which remains in the volume's I/O path. It manages the life cycle of the PV and PVCs while running in the user space. A variety of storage engines with varied capabilities are available on the data plane. **Jiva**, **cStor**, and **Local PV** are the three storage engines that are available at the time of writing. Jiva provides standard storage capabilities (block storage) and is typically used for smaller-scale workloads compared to cStor, which offers enterprise-grade functionality and extensive snapshot features. Local PV, on the other hand, provides performance through advanced features such as replication and snapshots.

Let's take a closer look at each storage engine.

Storage engines

OpenEBS's preferred storage engine is cStor. It's a feature-rich and lightweight storage engine designed for high-availability workloads such as databases. It includes enterprise-level capabilities such as synchronous data replication, snapshots, clones, thin data provisioning, high data resiliency, data consistency, and on-demand capacity or performance increases. With just a single replica, cStor's synchronous replication ensures excellent availability for stateful Kubernetes deployments. When a stateful application requires high data availability, cStor is set up with three replicas, with data written synchronously to each of the three replicas. Terminating and scheduling a new pod in a different node does not result in data loss because data is written to multiple replicas.

Jiva was the first storage engine to be included in early OpenEBS versions. Jiva is the simplest of the options, as it runs entirely in user space and has conventional block storage features such as synchronous replication. Smaller applications running on nodes without the ability to install extra block storage devices benefit from Jiva. As a result, it is not appropriate for mission-critical tasks that require high performance or advanced storage capacities.

Local persistent volume (**Local PV**) is OpenEBS's third and simplest storage engine. Local PV is a local disc that's attached directly to a single Kubernetes mode. Kubernetes applications can now consume high-performance local storage using the traditional volume APIs. OpenEBS's Local PV is a storage engine that may build PVs on worker nodes using local discs or host paths. Local PV can be used by cloud-native apps that do not require advanced storage features such as replication, snapshots, or clones. A StatefulSet that manages replication and HA on its own, for example, can set up a Local PV based on OpenEBS.

In addition to the storage engines mentioned previously, the **Mayastor data engine**, a low latency engine that is currently in development, has a declarative data plane, which provides flexible, persistent storage for stateful applications. It is Kubernetes-native and provides fast, redundant storage that works in any Kubernetes cluster. The Mayastor add-on will become available with MicroK8s 1.24: `https://microk8s.io/docs/addon-mayastor`.

Another optional and popular feature of OpenEBS is copy-on-write snapshots. Snapshots are created instantly, and there is no limit to the number of snapshots that can be created. The incremental snapshot feature improves data migration and portability across Kubernetes clusters, as well as between cloud providers or data centers. Common application scenarios include efficient replication for backups and the use of clones for troubleshooting or development against a read-only copy of data.

OpenEBS volumes also support backup and restore facilities that are compatible with Kubernetes backup and restore solutions such as Velero (`https://velero.io/`).

To learn more, you can check out my blog post on how to back up and restore Kubernetes cluster resources, including PVs: `https://www.upnxtblog.com/index.php/2019/12/16/how-to-back-up-and-restore-your-kubernetes-cluster-resources-and-persistent-volumes/`.

Through the container attached storage technique, OpenEBS extends the benefits of software-defined storage to cloud-native applications. For a thorough comparison and preferred use cases for each of the storage engines, see the OpenEBS documentation at `https://openebs.io/docs/`.

To recap, OpenEBS creates local or distributed Kubernetes PVs from any storage available to Kubernetes worker nodes. This makes it simple for application and platform teams to implement Kubernetes stateful workloads that require fast, reliable, and scalable CAS. It also ensures that each storage volume has a separate pod and a set of replica pods, which are managed and deployed in Kubernetes like any other container or microservice. OpenEBS is also installed as a container, allowing for convenient storage service allocation on a per-application, cluster, or container basis.

Now, let's learn how to configure and implement a PostgreSQL stateful application while utilizing the OpenEBS Jiva storage engine.

Configuring and implementing a PostgreSQL stateful workload

In this section, we'll configure and implement a PostgreSQL stateful workload while utilizing the OpenEBS storage engine. We'll be using the Jiva storage engine for PostgreSQL persistence, creating test data, simulating node failure to see if the data is still intact, and confirming that OpenEBS replication is functioning as expected.

Now that understand OpenEBS, we will delve into the steps of configuring and deploying OpenEBS on the cluster. The following diagram depicts our Raspberry Pi cluster setup:

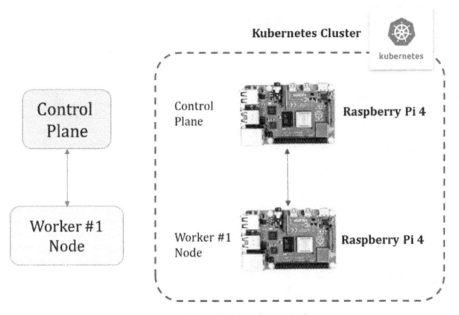

Figure 11.3 – MicroK8s Raspberry Pi cluster

Now that we know what we want to do, let's look at the requirements.

Requirements

Before you begin, you will need the following prerequisites to build a Raspberry Pi Kubernetes cluster and configure OpenEBS:

- A microSD card (4 GB minimum, 8 GB recommended)
- A computer with a microSD card drive
- A Raspberry Pi 2, 3, or 4 (1 or more)

- A micro-USB power cable (USB-C for the Pi 4)
- A Wi-Fi network or an ethernet cable with an internet connection
- (Optional) A monitor with an HDMI interface
- (Optional) An HDMI cable for the Pi 2 and 3 and a micro-HDMI cable for the Pi 4
- (Optional) A USB keyboard

Now that we've established what the requirements are for testing a PostgresSQL stateful workload backed by OpenEBS, let's get started.

Step 1 – Creating the MicroK8s Raspberry Pi cluster

Please follow the steps that we covered in *Chapter 5, Creating and Implementing Updates on Multi-Node Raspberry Pi Kubernetes Clusters*, to create the MicroK8s Raspberry Pi cluster; here's a quick refresher:

1. Install the OS image on the SD card:

    I.   Configure the Wi-Fi access settings.

    II. Configure the remote access settings.

    III. Configure the control group settings.

    IV. Configure the hostname.

2. Install and configure MicroK8s.

3. Add a worker node.

A fully functional multi-node Kubernetes cluster would look as follows. To summarize, we have installed MicroK8s on the Raspberry Pi boards and joined multiple deployments to form the cluster. We've also added nodes to the cluster:

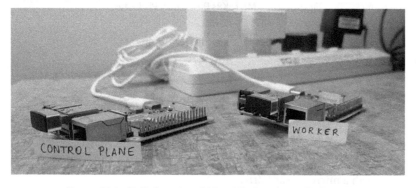

Figure 11.4 – Fully functional MicroK8s Raspberry Pi cluster

Now, let's enable the OpenEBS add-on.

Step 2 – Enabling the OpenEBS add-on

The OpenEBS add-on is available with MicroK8s by default. Use the following command to enable OpenEBS:

```
microk8s enable openebs
```

The output of the following command indicates that the `iscsid` controller must be enabled as a prerequisite. For storage management, OpenEBS uses the **Internet Small Computer System Interface (iSCSI)** technology. The iSCSI protocol is a TCP/IP-based protocol for creating storage area networks and establishing and managing interconnections between IP storage devices, hosts, and clients (SANs). These SANs allow the SCSI protocol to be used in high-speed data transmission networks with block-level data transfer between different data storage networks:

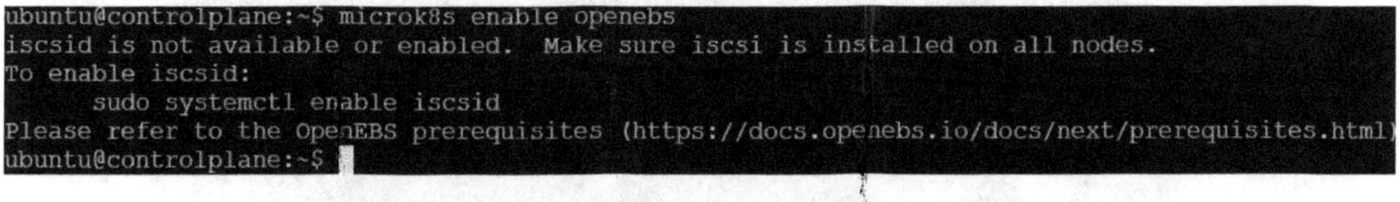

```
ubuntu@controlplane:~$ microk8s enable openebs
iscsid is not available or enabled.  Make sure iscsi is installed on all nodes.
To enable iscsid:
        sudo systemctl enable iscsid
Please refer to the OpenEBS prerequisites (https://docs.openebs.io/docs/next/prerequisites.html)
ubuntu@controlplane:~$
```

Figure 11.5 – Enabling the OpenEBS add-on

Use the following command to enable the `iscsid` controller:

```
sudo systemctl enable iscsid
```

The following output indicates that `iscsid` has been installed successfully. Now, we can enable the OpenEBS add-on:

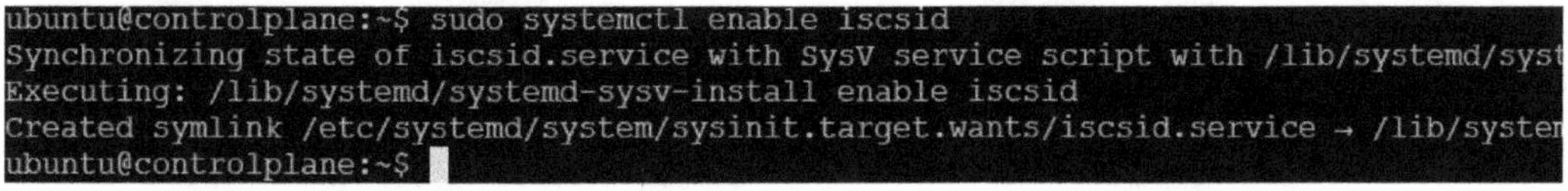

```
ubuntu@controlplane:~$ sudo systemctl enable iscsid
Synchronizing state of iscsid.service with SysV service script with /lib/systemd/syst
Executing: /lib/systemd/systemd-sysv-install enable iscsid
Created symlink /etc/systemd/system/sysinit.target.wants/iscsid.service → /lib/system
ubuntu@controlplane:~$
```

Figure 11.6 – Enabling the iSCSI controller

The following output indicates that the OpenEBS add-on has been enabled successfully:

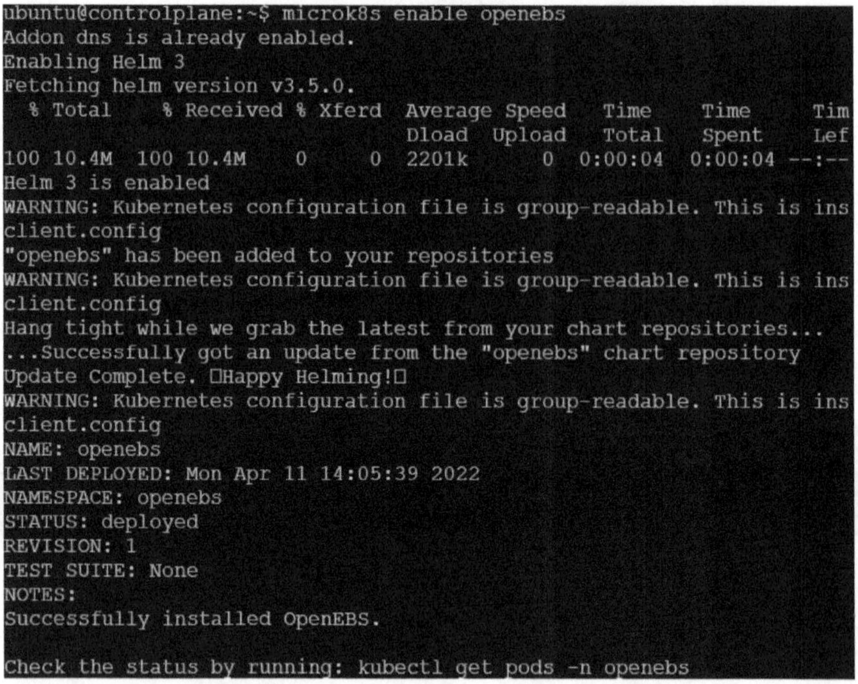

```
ubuntu@controlplane:~$ microk8s enable openebs
Addon dns is already enabled.
Enabling Helm 3
Fetching helm version v3.5.0.
  % Total    % Received % Xferd  Average Speed   Time    Time     Tim
                                 Dload  Upload   Total   Spent    Lef
100 10.4M  100 10.4M    0      0  2201k      0  0:00:04  0:00:04 --:--
Helm 3 is enabled
WARNING: Kubernetes configuration file is group-readable. This is ins
client.config
"openebs" has been added to your repositories
WARNING: Kubernetes configuration file is group-readable. This is ins
client.config
Hang tight while we grab the latest from your chart repositories...
...Successfully got an update from the "openebs" chart repository
Update Complete. □Happy Helming!□
WARNING: Kubernetes configuration file is group-readable. This is ins
client.config
NAME: openebs
LAST DEPLOYED: Mon Apr 11 14:05:39 2022
NAMESPACE: openebs
STATUS: deployed
REVISION: 1
TEST SUITE: None
NOTES:
Successfully installed OpenEBS.

Check the status by running: kubectl get pods -n openebs
```

Figure 11.7 – Enabling the OpenEBS add-on

The Helm3 add-on is also enabled by default. Before we move on, let's make sure that all of the OpenEBS components are up and running using the following command:

```
kubectl get pods -n openebs
```

The following output indicates that all the components are Running:

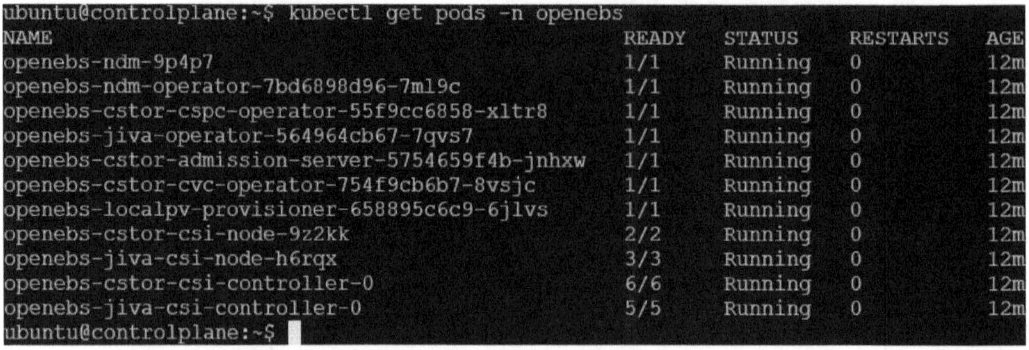

```
ubuntu@controlplane:~$ kubectl get pods -n openebs
NAME                                               READY   STATUS    RESTARTS   AGE
openebs-ndm-9p4p7                                  1/1     Running   0          12m
openebs-ndm-operator-7bd6898d96-7ml9c              1/1     Running   0          12m
openebs-cstor-cspc-operator-55f9cc6858-xltr8       1/1     Running   0          12m
openebs-jiva-operator-564964cb67-7qvs7             1/1     Running   0          12m
openebs-cstor-admission-server-5754659f4b-jnhxw    1/1     Running   0          12m
openebs-cstor-cvc-operator-754f9cb6b7-8vsjc        1/1     Running   0          12m
openebs-localpv-provisioner-658895c6c9-6jlvs       1/1     Running   0          12m
openebs-cstor-csi-node-9z2kk                       2/2     Running   0          12m
openebs-jiva-csi-node-h6rqx                        3/3     Running   0          12m
openebs-cstor-csi-controller-0                     6/6     Running   0          12m
openebs-jiva-csi-controller-0                      5/5     Running   0          12m
ubuntu@controlplane:~$
```

Figure 11.8 – The OpenEBS components are up and running

Now that the OpenEBS add-on has been enabled, let's deploy a PostgreSQL stateful workload.

Step 3 – Deploying the PostgreSQL stateful workload

To recap from *Chapter 1*, *Getting Started with Kubernetes*, a StatefulSet is a Kubernetes workload API object for managing stateful applications. In a typical deployment, the user is not concerned about how the pods are scheduled, so long as it has no negative impact on the deployed application. However, to preserve the state in stateful applications with persistent storage, pods must be identified. This functionality is provided by StatefulSet, which creates pods with a persistent identifier that corresponds to its value across rescheduling. This way, even if a pod is recreated, it will be correctly mapped to the storage volumes, and the application's state will be preserved.

With the popularity of deploying database clusters in Kubernetes, managing states in a containerized environment have become even more important.

We'll need to set up the following resources to get the PostgreSQL configuration up and running:

- Storage class
- PersistentVolume
- PersistentVolumeClaim
- StatefulSet
- ConfigMap
- Service

To manage persistent storage, Kubernetes provides the `PersistentVolume` and `PersistentVolumeClaim` storage mechanisms, which we briefly discussed in the introduction. Here's a quick rundown of what they are:

- **PersistentVolume (PV)** is stored in a cluster that has been provisioned by a cluster administrator or dynamically provisioned using storage classes.

- **PersistentVolumeClaim (PVC)** is a user's (developer's) request for storage. It is comparable to a pod. PVCs consume PV resources, while pods consume node resources:

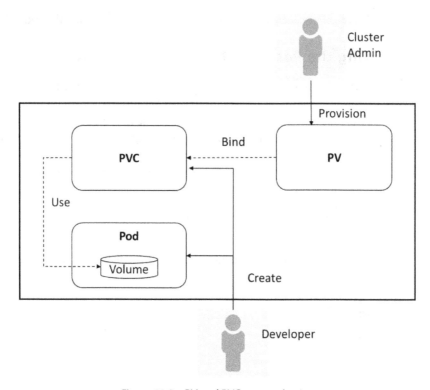

Figure 11.9 – PV and PVC storage basics

Before we create a PV and PVC, let's look at the storage class that OpenEBS has created for us.

StorageClass allows administrators to describe the *classes* of storage that they provide. Different classes may correspond to different **Quality-of-Service (QoS)** levels, backup policies, or arbitrary policies that are determined by the cluster administrators.

Use the following command to retrieve the storage class that has been created by OpenEBS:

```
kubectl get sc
```

The following output shows that three StorageClass are available:

```
ubuntu@controlplane:~$ kubectl get sc
NAME                      PROVISIONER          RECLAIMPOLICY   VOLUMEBINDINGMODE     ALLOWVO
openebs-jiva-csi-default  jiva.csi.openebs.io  Delete          Immediate             true
openebs-device            openebs.io/local     Delete          WaitForFirstConsumer  false
openebs-hostpath          openebs.io/local     Delete          WaitForFirstConsumer  false
ubuntu@controlplane:~$ 
```

Figure 11.10 – OpenEBS storage classes

`openebs-hostpath` and `openebs-device` are recommended for single-node clusters. For multi-node clusters, `openebs-jiva-csi-default` is recommended.

Now, we must define `PersistentVolume`, which will use the storage class, as well as `PersistentVolumeClaim`, which will be used to claim this volume:

```
kind: PersistentVolume
apiVersion: v1
metadata:
  name: postgres-pv
  labels:
    app: postgres
    type: local
spec:
  storageClassName: openebs-jiva-csi-default
  capacity:
    storage: 5Gi
  accessModes:
    - ReadWriteOnce
  hostPath:
    path: "/var/data"
---
kind: PersistentVolumeClaim
apiVersion: v1
metadata:
  name: postgres-pv-claim
  labels:
    app: postgres
spec:
  storageClassName: openebs-jiva-csi-default
  capacity:
  accessModes:
    - ReadWriteOnce
  resources:
    requests:
      storage: 5Gi
```

Because we're utilizing an OpenEBS disc provisioner, we'll need to specify where our data will be saved on the host node. We'll use `/var/data/` in this case. The `accessMode` option is also crucial. We'll use `ReadWriteOnce` in this case. This ensures that only one pod can write at any given moment. As a result, no two pods end up with the same writing volume. We can also specify the size of this volume, which we chose to be 5 GB.

> **Note on Access Modes**
>
> Even though a volume supports several access modes, they can only be mounted one at a time:
>
> **ReadOnlyMany (ROX)**: Can be mounted by multiple nodes in read-only mode
>
> **ReadWriteOnce (RWO)**: Can be mounted by a single node in read-write mode
>
> **ReadWriteMany (RWX)**: Multiple nodes can be mounted in read-write mode

Use the following command to create the PV and PVC:

```
kubectl apply -f postgres.yaml
```

The following output indicates that `PersistentVolume` and `PersistentVolumeClaim` have been created successfully:

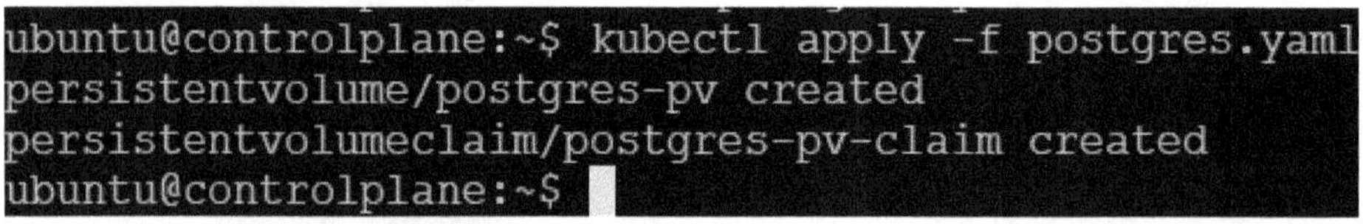

```
ubuntu@controlplane:~$ kubectl apply -f postgres.yaml
persistentvolume/postgres-pv created
persistentvolumeclaim/postgres-pv-claim created
ubuntu@controlplane:~$
```

Figure 11.11 – PV and PVC created successfully

Before moving on, let's check if PV and PVC are Bound. A Bound state indicates that the application has access to the necessary storage:

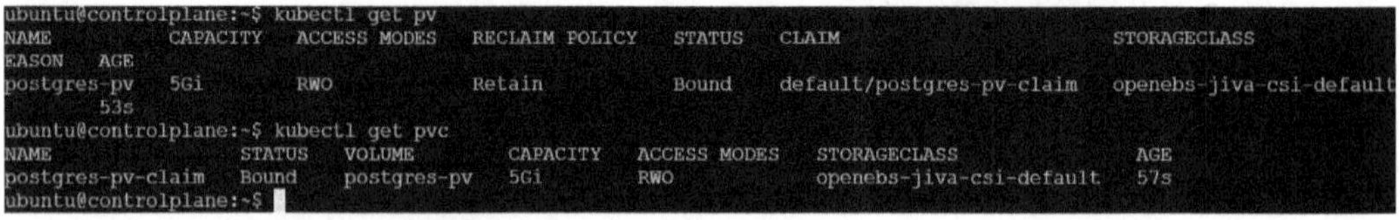

```
ubuntu@controlplane:~$ kubectl get pv
NAME          CAPACITY   ACCESS MODES   RECLAIM POLICY   STATUS   CLAIM                       STORAGECLASS
EASON   AGE
postgres-pv   5Gi        RWO            Retain           Bound    default/postgres-pv-claim   openebs-jiva-csi-default
        53s
ubuntu@controlplane:~$ kubectl get pvc
NAME                STATUS   VOLUME        CAPACITY   ACCESS MODES   STORAGECLASS               AGE
postgres-pv-claim   Bound    postgres-pv   5Gi        RWO            openebs-jiva-csi-default   57s
ubuntu@controlplane:~$
```

Figure 11.12 – PV and PVC are bound

If a PVC becomes stuck waiting, `StatefulSet` will get stuck as well, as it will be unable to access its storage. As a result, double-check that both `StorageClass` and `PersistentVolume` have been set up correctly.

Now that we've set up the PV and PVC, we'll set up `ConfigMap` with configurations such as the username and password required for our setup. To keep things simple in this example, we've hardcoded the values inside `ConfigMap`:

```
apiVersion: v1
kind: ConfigMap
metadata:
  name: postgres-configuration
  labels:
    app: postgres
data:
  POSTGRES_DB: postgresdb
  POSTGRES_USER: postgres
  POSTGRES_PASSWORD: postgrespassword
```

Use the following command to create `ConfigMap`:

```
kubectl apply -f postgres-config.yaml
```

The following output indicates that the `postgres-configuration.yaml` file's `ConfigMap` has been created successfully:

```
ubuntu@controlplane:~$ kubectl apply -f postgres-config.yaml
configmap/postgres-configuration created
ubuntu@controlplane:~$
```

Figure 11.13 – PostgreSQL ConfigMap created

Let's use the `describe` command to fetch the details of the `ConfigMap` object that we have created. The following output shows that the configuration required for our PostgreSQL setup is ready:

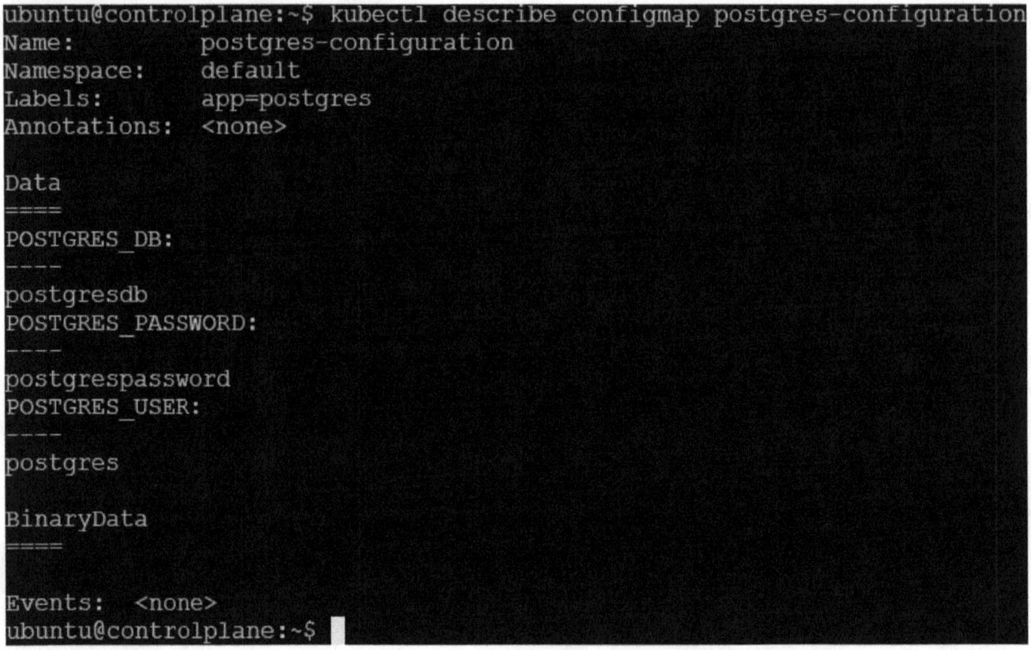

```
ubuntu@controlplane:~$ kubectl describe configmap postgres-configuration
Name:          postgres-configuration
Namespace:     default
Labels:        app=postgres
Annotations:   <none>

Data
====
POSTGRES_DB:
----
postgresdb
POSTGRES_PASSWORD:
----
postgrespassword
POSTGRES_USER:
----
postgres

BinaryData
====

Events:   <none>
ubuntu@controlplane:~$
```

Figure 11.14 – PostgreSQL ConfigMap

Now that we've defined our `ConfigMap` and storage volume, we can define a `StatefulSet` that will make use of them:

```
apiVersion: apps/v1
kind: StatefulSet
metadata:
  name: postgres-statefulset
  labels:
    app: postgres
spec:
  serviceName: "postgres"
  replicas: 2
  selector:
    matchLabels:
      app: postgres
  template:
    metadata:
```

```
      labels:
        app: postgres
    spec:
      containers:
      - name: postgres
        image: postgres:12
        envFrom:
        - configMapRef:
            name: postgres-configuration
        ports:
        - containerPort: 5432
          name: postgresdb
        volumeMounts:
        - name: pv-data
          mountPath: /var/lib/postgresql/data
      volumes:
      - name: pv-data
        persistentVolumeClaim:
          claimName: postgres-pv-claim
---
apiVersion: v1
kind: Service
metadata:
  name: postgres-service
  labels:
    app: postgres
spec:
  ports:
  - port: 5432
    name: postgres
  type: NodePort
  selector:
    app: postgres
```

The definition of a `StatefulSet` is similar to that of deployments. We've added two more things:

- We've loaded the environment variables from `ConfigMap` into the pod.
- We've defined our volume, which will map to `/var/lib/PostgreSQL/data` within our pod. This volume is defined using the PVC that we discussed earlier.

Finally, we have also created a `Service` resource that will expose our database.

Use the following command to create the `StatefulSet` and `Service` resources to expose the database:

```
kubectl apply -f postgres-deployment.yaml
```

The following output indicates that both `StatefulSet` and `Service` have been created successfully:

```
ubuntu@controlplane:~$ kubectl apply -f postgres-deployment.yaml
statefulset.apps/postgres-statefulset created
service/postgres-service created
ubuntu@controlplane:~$
```

Figure 11.15 – Postgres deployment succeeded

Before moving on, let's verify that the pods and service have been created. The following output shows that the pods are `Running`:

```
ubuntu@controlplane:~$ kubectl get pods | grep post
postgres-statefulset-0    1/1       Running    0              3m22s
ubuntu@controlplane:~$
```

Figure 11.16 – The Postgres pods are Running

The following output shows that the service has been exposed on port `5432`:

```
ubuntu@controlplane:~$ kubectl get svc | grep post
postgres-service          NodePort      10.152.183.6      <none>          5432:32367/TCP    3m50s
ubuntu@controlplane:~$
```

Figure 11.17 – The Postgres service has been exposed

Let's also look at where the `StatefulSet` pods are distributed across the cluster using the following command:

```
kubectl get pods -o wide | grep post
```

The following output shows that the PostgreSQL database pods are running on two nodes (1 in `controlplane` and 1 in `worker1`):

```
ubuntu@controlplane:~$ kubectl get pods -o wide | grep post
postgres-statefulset-0   1/1      Running   0          16m    10.1.49.81      controlplane   <none>
postgres-statefulset-1   1/1      Running   0          107s   10.1.235.151   worker1        <none>
ubuntu@controlplane:~$
```

Figure 11.18 – The database pods are running on two nodes

With that, we have successfully configured PostgreSQL and it's up and running.

Now, let's create a test database and a table, and add a few records.

Step 4 – Creating the test data

To create test data, use the `PgSQL` client or log into one of the pods so that we can create a test database and table.

Use the following command to log into the PostgreSQL pod:

```
kubectl exec -it postgres-statefulset-0 -- psql -U postgres
```

The following output shows that we can log into the PostgreSQL pod:

```
ubuntu@controlplane:~$ kubectl exec -it postgres-statefulset-0 -- psql -U postgres
psql (12.10 (Debian 12.10-1.pgdg110+1))
Type "help" for help.

postgres=#
```

Figure 11.19 – Logging into one of the PostgreSQL pods

Now that we have logged into the pod, we have access to the `psql` PostgreSQL client. Use the following command to create the test database:

```
CREATE DATABASE inventory_mgmt;
```

The following output shows that our test database, `inventory_mgmt`, has been created:

```
postgres=# CREATE DATABASE inventory_mgmt;
CREATE DATABASE
postgres=#
```

Figure 11.20 – Creating the test database

Let's switch our connection to the new database we have created using \c inventory_mgmt. The following output indicates that we have successfully switched to a new database. Now, we can create a table:

```
postgres-# \c inventory_mgmt
You are now connected to database "inventory_mgmt" as user "postgres".
inventory_mgmt-#
```

Figure 11.21 – Switching the connection to the new database

In the new database, use the CREATE TABLE command to create a test table. The following output indicates that a new table called products_master has been created successfully:

```
inventory_mgmt=# CREATE TABLE products_master (
    product_no integer,
    name text,
    price numeric
);
CREATE TABLE
inventory_mgmt=#
```

Figure 11.22 – Creating the test table

Now that the test table has been created, use the INSERT command to add a few records, as shown here:

```
inventory_mgmt=# INSERT INTO products_master VALUES (1, 'Chair', 119.99);
INSERT 0 1
inventory_mgmt=#
inventory_mgmt=# INSERT INTO products_master VALUES (2, 'Work Table', 199.99);
INSERT 0 1
inventory_mgmt=#
```

Figure 11.23 – Adding a few records to the test table

Here, we have added records to our test table. Before we move on, let's use the SELECT command to list the records, as shown in the following screenshot:

```
inventory_mgmt=# SELECT * FROM products_master;
 product_no |     name     | price
------------+--------------+--------
          1 | Chair        | 119.99
          2 | Work Table   | 199.99
(2 rows)

inventory_mgmt=# 
```

Figure 11.24 – Records from the test table

To recap, in this section, we have created a test database, created a new table, and added a few records to the table. Now, let's simulate node failure.

Step 5 – Simulating node failure

To simulate node failure, we will use the `cordon` command to mark the node as `unschedulable`. If the node is `unschedulable`, the Kubernetes controller will not schedule new pods on this node.

Let's locate the PostgreSQL database pod's node and cordon it off, preventing new pods from being scheduled on it.

The following output shows that the database pods are running on 2 nodes (1 in `controlplane` and 1 in `worker1`):

```
ubuntu@controlplane:~$ kubectl get pods -o wide | grep post
postgres-statefulset-0   1/1     Running   0          16m     10.1.49.81      controlplane   <none>
postgres-statefulset-1   1/1     Running   0          107s    10.1.235.151    worker1        <none>
ubuntu@controlplane:~$ 
```

Figure 11.25 – PostgreSQL database pods

Let's use `cordon` on the `worker1` node so that new pods are prevented from being scheduled on it:

```
kubectl cordon worker1
```

The following output shows that `worker1` has been cordoned:

```
ubuntu@controlplane:~$ kubectl cordon worker1
node/worker1 cordoned
ubuntu@controlplane:~$ 
```

Figure 11.26 – Cordoned Worker1 node

Even though the `worker1` node has been cordoned, existing pods will still run, so we can use the `drain` command to delete all the pods:

```
kubectl drain --force --ignore-daemonsets worker1
```

The following output shows that `worker1` can't be drained due to pods with local storage provisioned:

```
ubuntu@controlplane:~$ kubectl drain --force --ignore-daemonsets worker1
node/worker1 already cordoned
error: unable to drain node "worker1" due to error:cannot delete Pods with local storage (use
erride): openebs/openebs-jiva-csi-controller-0, continuing command...
There are pending nodes to be drained:
 worker1
cannot delete Pods with local storage (use --delete-emptydir-data to override): openebs/opene
```

Figure 11.27 – Draining the Worker1 node

Finally, we will use the `kubectl delete` command to delete the pod that is currently running on the cordoned node.

The following output shows that pods running on `worker1` have been deleted successfully:

```
ubuntu@controlplane:~$ kubectl delete pod postgres-statefulset-1
pod "postgres-statefulset-1" deleted
ubuntu@controlplane:~$
```

Figure 11.28 – Deleting the pods running on the Worker1 node

The Kubernetes controller will now recreate a new pod and schedule it in a different node as soon as the pod is deleted. It cannot be placed on the same node since scheduling has been disabled; this is because we cordoned the `worker1` node.

Let's inspect where the pods are running using the `kubectl get pods` command. The following output shows that the new pod has been rescheduled to the `controlplane` node:

```
ubuntu@controlplane:~$ kubectl get pods -o wide | grep post
postgres-statefulset-0    1/1    Running    0    24m    10.1.49.81    controlplane    <none>    <none>
postgres-statefulset-1    1/1    Running    0    13s    10.1.49.69    controlplane    <none>    <none>
ubuntu@controlplane:~$
```

Figure 11.29 – PostgreSQL database pods

Even though the PVC has a `ReadWriteOnce` access mode and is mounted by a specific node for read-write access, the new pod that has been recreated can use the same PVC that has been abstracted by the underlying `OpenEBS` volumes into a single storage layer.

To verify if the new pod is using the same PVC, let's connect to the new pod and see if the data is still intact by using the `kubectl exec` command:

```
ubuntu@controlplane:~$ kubectl exec -it postgres-statefulset-1 -- psql -U postgres
psql (12.10 (Debian 12.10-1.pgdg110+1))
Type "help" for help.

postgres=#
```

Figure 11.30 – Logging into the PostgreSQL pod

The following output shows that the data is intact even after deleting the pod and rescheduling it on a different node. This confirms that the replication `OpenEBS` is working properly:

```
ubuntu@controlplane:~$ kubectl exec -it postgres-statefulset-1 -- psql -U postgres
psql (12.10 (Debian 12.10-1.pgdg110+1))
Type "help" for help.

postgres=# \c inventory_mgmt;
You are now connected to database "inventory_mgmt" as user "postgres".
inventory_mgmt=# select * from products_master;
 product_no |    name     | price
------------+-------------+-------
          1 | Chair       | 119.99
          2 | Work Table  | 199.99
(2 rows)

inventory_mgmt=#
```

Figure 11.31 – The data is intact

To summarize, data engines are responsible for maintaining the actual state generated by stateful applications, as well as providing sufficient storage capacity to retain the information and ensure that it remains intact over time. For example, the state can be created once, accessed over the next few minutes or days, updated, or simply left to be retrieved months or years later. You can use **Local PV**, **Jiva**, **cStor**, or **Mayastor**, depending on the type of storage associated with your Kubernetes worker nodes and your application performance needs.

Choosing an engine is entirely dependent on your platform (resources and storage type), the application workload, and the application's current and future capacity and/or performance growth. In the next section, we'll look at some Kubernetes storage best practices, as well as some recommendations for data engines.

Kubernetes storage best practices

For modern containerized applications deployed on Kubernetes, storage is a crucial concern. Kubernetes has progressed from local node filesystems mounted in containers to NFS, and finally to native storage, as described by the CSI specification, which allows for data durability and sharing. In this section, we'll look at some of the best practices to take into consideration when configuring a PV:

- Avoid statically creating and allocating PVs to decrease management costs and facilitate scaling. Use dynamic provisioning instead. Define an appropriate reclaim policy in your storage class to reduce storage costs when pods are deleted.

- Each node can only support a certain number of sizes, so different node sizes provide varying amounts of local storage and capacity. To install the optimum node sizes, plan accordingly for your application's demands.

- The life cycle of a PV is independent of any individual container in the cluster. A PV is a request for a specific type of storage made by a container user or application. Kubernetes documentation suggests the following for building a PV:

  - PVCs should always be included in the container setup.

  - PVs should never be used in container configuration since they will bind a container to a specific volume.

  - PVCs that don't specify a specific class will fail if they don't have a default `Storage Class`.

  - Give Storage Classes names that are meaningful.

- At the namespace level, resource quotas are also provided, giving you another level of control over cluster resource utilization. The total amount of CPU, memory, and storage resources that all the containers executing in the namespace can utilize is limited by resource limits. It can also set storage resource limits based on service levels or backup requirements.

- Persistent storage hardware comes in a variety of shapes and sizes. SSDs, for example, outperform HDDs in terms of read/write performance, and NVMe SSDs are especially well-suited to high workloads. QoS criteria are added to the description of a PVC by some of the Kubernetes providers. This means that it prioritizes read/write volumes for specific installations, allowing for higher performance if the application requires it.

Now, let's look at some of the guidelines for selecting OpenEBS data engines.

Guidelines on choosing OpenEBS data engines

Each storage engine has its advantages, as shown in the following table. Choosing an engine is entirely dependent on your platform (resources and storage type), the application workload, and the application's current and future capacity and/or performance growth. The following guidelines will assist you in selecting an engine:

Parameter	Local PV	JIVA	CSTOR	MAYASTOR
Support for full backup and restore	Yes	Yes	Yes	Yes
Synchronous replication support	N/A	Yes	Yes	Yes
Protect against node failures (replace node)	N/A	Yes	Yes	Yes
Use with ephemeral storage on nodes	N/A	Yes	Yes	Yes
Support for thin provisioning to create volumes with a size that is substantially larger than the available storage	Yes	Yes	Yes	Yes
Disk pool or aggregate support	Yes	Yes	Yes	Planned
Disk resiliency (RAID support)	Yes	Yes	Yes	Planned
On-demand capacity expansion	Yes	No	Yes	Planned
Support for snapshots	No	No	Yes	Planned
Support for clones	No	No	Yes	Planned
Support for incremental backups	No	No	Yes	Planned
Suitable for high capacity (>50 GB) workloads	No	No	Yes	Yes
Near-disk performance	Yes	No	No	Yes

Table 11.1 – Choosing OpenEBS data engines

In conclusion, OpenEBS offers a set of data engines, each of which is designed and optimized for executing stateful workloads with varied capabilities on Kubernetes nodes with varying resource levels. In a Kubernetes cluster, platform SREs or administrators often choose one or more data engines. These data engines are chosen based on node capabilities or stateful application capabilities.

Summary

In this chapter, we learned how Kubernetes persistent storage provides a convenient way for Kubernetes applications to request and consume storage resources. The PVC is declared by the user's pod, and Kubernetes will find a PV to pair it with. If there is no PV to pair with, then it will go to the corresponding `StorageClass` and assist it in creating a PV before binding it to the PVC. The newly created PV must use the attached master node to create a remote disc for the host and then mount the attached remote disc to the host directory using the `kubelet` component of each node.

Kubernetes has made significant improvements to facilitate running stateful workloads by giving platform (or cluster administrators) and application developers the necessary abstractions. These abstractions ensure that different types of file and block storage (whether ephemeral or persistent, local or remote) are available wherever a container is scheduled (including provisioning/creating, attaching, mounting, unmounting, detaching, and deleting volumes), storage capacity management (container ephemeral storage usage, volume resizing, and generic operations), and influencing container scheduling based on storage (data gravity, availability, and so on).

In the next chapter, you will learn how to deploy the Istio and Linkerd service mesh. You will also learn how to deploy and run a sample application, as well as how to configure and access dashboards.

12

Implementing Service Mesh for Cross-Cutting Concerns

In the previous chapter, we looked at OpenEBS cloud-native storage solutions so that we can provide persistent storage for our container applications. We also looked at how **Container Attached Storage (CAS)** is swiftly gaining acceptance as a viable solution for managing stateful workloads and utilizing persistent, fault-tolerant stateful applications. MicroK8s comes with built-in support for OpenEBS, making it the ideal option for running Kubernetes clusters in air-gapped Edge/IoT environments. Using the OpenEBS storage engine, we configured and implemented a PostgreSQL stateful workload. We also went over some best practices to keep in mind when creating a persistent volume and while selecting OpenEBS data engines.

The emergence of cloud-native applications is linked to the rise of the service mesh. In the cloud-native world, an application could be made up of hundreds of services, each of which could have thousands of instances, each of which could be constantly changing due to an orchestrator such as Kubernetes dynamically scheduling them. Not only is service-to-service communication tremendously complex, but it's also a critical component of an application's runtime behavior. It is critical to manage it to ensure end-to-end performance, dependability, and security.

A service mesh, such as Linkerd or Istio, is a tool for transparently embedding observability, security, and reliability features into cloud-native applications at the infrastructure layer rather than the application layer. The service mesh is quickly becoming an essential component of the cloud-native stack, particularly among Kubernetes users. Typically, the service mesh is built as a scalable set of network proxies that run alongside application code (the sidecar pattern). These proxies mediate communication between microservices and serve as a point where service mesh functionalities can be implemented.

The service mesh layer can run in a container alongside the application as a sidecar. Each of the applications has many copies of the same sidecar attached to it, as shown in the following diagram:

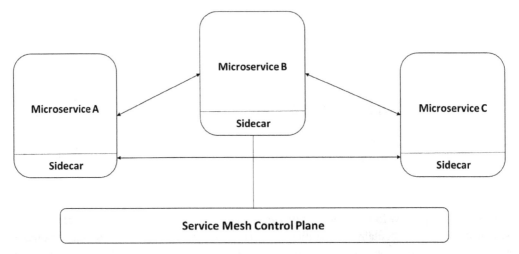

Figure 12.1 – The service mesh sidecar pattern

The sidecar proxy handles all incoming and outgoing network traffic from a single service. As a result, the sidecar controls traffic between microservices, collects telemetry data, and applies policies. In certain ways, the application service is unaware of the network and is just aware of the sidecar proxy connecting to it.

Within a service mesh, there's a data plane and a control plane:

- The **data plane** coordinates communication between the mesh's services and performs functions such as service discovery, load balancing, traffic management, health checks, and so on.

- The **control plane** manages and configures sidecar proxies so that policies can be enforced and telemetry can be collected.

The service mesh provides features for service discovery, automatic load balancing, fine-grained control of traffic behavior with routing rules, retries, failovers, and more. It also has a pluggable policy layer and API configuration that supports access controls, rate limits, and quotas. Finally, it provides service monitoring with automatic metrics, logs, and traces for all traffic, as well as secure service-to-service communication in the mesh.

In this chapter, we will look at two popular service mesh providers to implement this pattern: Linkerd and Istio. We won't be looking at all the capabilities of a service mesh; instead, we will touch upon the monitoring aspect using a sample application.

In this chapter, we're going to cover the following topics:

- Overview of the Linkerd service mesh

- Enabling the Linkerd add-on and running a sample application

- Overview of the Istio service mesh

- Enabling the Istio add-on and running a sample application

- Common use cases for a service mesh

- Guidelines on choosing a service mesh

- Best practices for configuring a service mesh

Overview of the Linkerd service mesh

Linkerd is a Kubernetes-based service mesh. It simplifies and secures how services operate by providing runtime debugging, observability, dependability, and security all without requiring any code changes.

Each service instance is connected to Linkerd by a system of ultralight, transparent proxies. These proxies handle all traffic to and from the service automatically. These proxies function as highly instrumented out-of-process network stacks, sending telemetry to and receiving control signals from the control plane. Linkerd can also measure and manage traffic to and from the service without introducing unnecessary latency.

As discussed in the previous chapter, Linkerd is made up of a control plane, which is a collection of services that control Linkerd as a whole, and a data plane, which is made up of transparent micro-proxies that run closer to each service instance in the pods as sidecar containers. These proxies handle all TCP traffic to and from the service automatically and communicate with the control plane for configuration.

The following diagram shows the architecture of Linkerd:

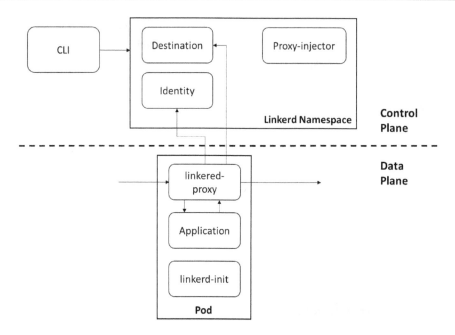

Figure 12.2 – Linkerd service mesh components

Now that we've provided a high-level overview and looked at the architecture, let's look at each component in more detail:

- **Destination service**: The data plane proxies use the destination service to determine various aspects of their behavior. It is used to retrieve service discovery information, retrieve policy information about which types of requests are permitted, and retrieve service profile information that's used to inform per-route metrics, retries, timeouts, and more.

- **Identity service**: The identity service functions as a TLS Certificate Authority, accepting CSRs from proxies and issuing signed certificates. These certificates are issued at proxy initialization time and are used to implement mTLS on proxy-to-proxy connections.

- **Proxy injector**: Every time a pod is created, the proxy injector receives a webhook request from Kubernetes. This injector looks for a Linkerd-specific annotation (`linkerd.io/inject: enabled`) in resources. When that annotation is present, the injector modifies the pod's specification and adds the `proxy-init` and `linkerd-proxy` containers, as well as the relevant start-time configuration, to the pod.

- **Linkerd2-proxy**: Linkerd2-proxy is an ultralight, transparent micro-proxy that was created specifically for the service mesh use case and is not intended to be a general-purpose proxy.

- **Linkerd-init container**: Each meshed pod has the `linkerd-init` container added as a Kubernetes `init` container that runs before any other containers are started. All TCP traffic to and from the pod is routed through the proxy using iptables.

Now that we've grasped the fundamentals, let's enable the Linkerd add-on and run a sample application.

Enabling the Linkerd add-on and running a sample application

In this section, you will enable the Linkerd add-on in your MicroK8s Kubernetes cluster. Then, to demonstrate Linkerd's capabilities, you will deploy a sample application.

> **Note**
>
> I'll be using an Ubuntu virtual machine for this section. The instructions for setting up a MicroK8s cluster are the same as those in *Chapter 5, Creating and Implementing Updates on Multi-Node Raspberry Pi Kubernetes Clusters*.

Step 1 – Enabling the Linkerd add-on

Use the following command to enable the Cilium add-on:

```
microk8s enable linkerd
```

The following output indicates that the Linkerd add-on has been enabled:

```
$ microk8s enable linkerd
Fetching Linkerd2 version v2.11.1.
2.11.1
  % Total    % Received % Xferd  Average Speed   Time    Time     Time  Current
                                 Dload  Upload   Total   Spent    Left  Speed
100   677  100   677    0     0   2763      0 --:--:-- --:--:-- --:--:--  2763
100 48.6M  100 48.6M    0     0  18.1M      0  0:00:02  0:00:02 --:--:-- 22.6M
Enabling Linkerd2
Enabling DNS
Applying manifest
serviceaccount/coredns created
configmap/coredns created
deployment.apps/coredns created
service/kube-dns created
clusterrole.rbac.authorization.k8s.io/coredns created
clusterrolebinding.rbac.authorization.k8s.io/coredns created
Restarting kubelet
DNS is enabled
namespace/linkerd created
```

Figure 12.3 – Enabling the Linkerd add-on

It will take some time to finish activating the add-on. The following output shows that Linkerd has been successfully enabled:

```
mutatingwebhookconfiguration.admissionregistration.k8s.io/linkerd-tap-injec
service/tap-injector created
deployment.apps/tap-injector created
server.policy.linkerd.io/tap-injector-webhook created
serverauthorization.policy.linkerd.io/tap-injector created
service/web created
deployment.apps/web created
serviceprofile.linkerd.io/metrics-api.linkerd-viz.svc.cluster.local created
serviceprofile.linkerd.io/prometheus.linkerd-viz.svc.cluster.local created
serviceprofile.linkerd.io/grafana.linkerd-viz.svc.cluster.local created
Linkerd is starting
$
```

Figure 12.4 – Linkerd enabled successfully

Before we move on to the next step, let's make sure that all of the Linkerd components are up and running by using the following command:

```
kubectl get pods -n linkerd
```

The following output indicates that all the components are Running:

```
$ kubectl get pods -n linkerd
NAME                                         READY   STATUS
linkerd-identity-54795b9f9f-gh9cl            2/2     Running
linkerd-destination-8468749f6f-4jv5f         4/4     Running
linkerd-proxy-injector-6db5c59479-hr4qw      2/2     Running
$
```

Figure 12.5 – The Linkerd pods are running

Now that the Linkerd add-on has been enabled, let's deploy a sample Nginx application.

Step 2 – Deploying the sample application

In this step, we will be deploying a sample Nginx application from the Kubernetes sample repository.

Use the following command to create a sample Nginx deployment:

```
kubectl apply -f https://k8s.io/examples/application/
deployment.yaml
```

The following output indicates that there is no error in the deployment. Now, we can ensure that the pods have been created:

```
$ kubectl apply -f  https://k8s.io/examples/application/deployment.yaml
deployment.apps/nginx-deployment created
$ ▮
```

Figure 12.6 – Sample Nginx application deployment

Now that the deployment is successful, let's use the `kubectl` command to check if the pods are `Running`:

```
$ kubectl get pods
NAME                                     READY     STATUS
nginx-deployment-9456bbbf9-8v4s8         1/1       Running
nginx-deployment-9456bbbf9-cm59w         1/1       Running
$ ▮
```

Figure 12.7 – Sample application pods

Here, we can see that the sample application deployment has been successful and that all the pods are running. Our next step is to inject Linkerd into it by piping the `linkerd inject` and `kubectl apply` commands together. Without any downtime, Kubernetes will perform a rolling deployment and update each pod with the data plane's proxies.

Use the following command to inject Linkerd into the sample application:

```
kubectl get deployment nginx-deployment -n default -o yaml |
microk8s linkerd inject - | kubectl apply -f -
```

The following output confirms that the `linkerd inject` command has succeeded and that the sample application deployment has been reconfigured:

```
$    kubectl get deployment nginx-deployment -n default -o yaml \
>    | microk8s linkerd inject - \
>    | kubectl apply -f -

deployment "nginx-deployment" injected

deployment.apps/nginx-deployment configured
$ ▮
```

Figure 12.8 – Injecting Linkerd into the sample application

The `linkerd inject` command simply adds annotations to the pod spec instructing Linkerd to inject the proxy into pods when they are created.

Congratulations! Linkerd has now been added to the sample Nginx application! We added Linkerd to sample application services without touching the original YAML.

On the data plane side, it's possible to double-check that everything is working properly. Examine the data plane with the following command:

```
microk8s linkerd check --proxy
```

The Linkerd CLI (`microk8s linkerd`) is the main interface for working with Linkerd. It can set up the control plane on the cluster, add the proxy to the service(s), and offer thorough performance metrics for the service(s).

The following output confirms that the `linkerd check` command has started the checks for the data plane:

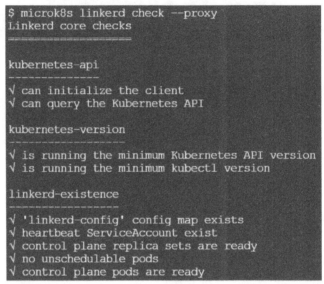

Figure 12.9 – Linkerd checks

It will take some time to finish the data plane checks. The following output shows that the Linkerd checks have been completed:

```
linkerd-viz
-----------
√ linkerd-viz Namespace exists
√ linkerd-viz ClusterRoles exist
√ linkerd-viz ClusterRoleBindings exist
√ tap API server has valid cert
√ tap API server cert is valid for at least 60
√ tap API service is running
√ linkerd-viz pods are injected
√ viz extension pods are running
√ viz extension proxies are healthy
√ viz extension proxies are up-to-date
√ viz extension proxies and cli versions match
√ prometheus is installed and configured correc
√ can initialize the client
√ viz extension self-check

Status check results are √
$
```

Figure 12.10 – Linkerd checks completed

Now that the data plane checks have been completed, we can see if the Linkerd annotations have been added to the sample application deployment by using the kubectl describe command.

The following output confirms that Linkerd annotations have been added:

```
$ kubectl describe deployment nginx-deployment
Name:                     nginx-deployment
Namespace:                default
CreationTimestamp:        Sun, 17 Apr 2022 12:05:28 +0000
Labels:                   <none>
Annotations:              deployment.kubernetes.io/revision: 2
Selector:                 app=nginx
Replicas:                 2 desired | 2 updated | 2 total | 2
StrategyType:             RollingUpdate
MinReadySeconds:          0
RollingUpdateStrategy:    25% max unavailable, 25% max surge
Pod Template:
  Labels:        app=nginx
  Annotations:   linkerd.io/inject: enabled
  Containers:
   nginx:
    Image:          nginx:1.14.2
    Port:           80/TCP
    Host Port:      0/TCP
    Environment:    <none>
    Mounts:         <none>
```

Figure 12.11 – Linkerd annotations added

Furthermore, we have injected Linkerd without having to write any special configurations or change the code of the application. If we can provide Linkerd with additional information, it will be able to impose a variety of restrictions, such as timeouts and retries. Then, it can provide stats for each route.

Next, we will start retrieving vital information about how each of the services of the sample Nginx deployment is performing. Since Linkerd has been injected into the application, we will look at various metrics and dashboards

Step 3 – Exploring the Linkerd dashboard

Linkerd offers an on-cluster metrics stack that includes a web dashboard and pre-configured Grafana dashboards. In this step, we will learn how to launch the Linkerd and Grafana dashboards.

Use the following command to launch the Linkerd dashboard:

```
microk8s linkerd viz dashboard
```

The following output indicates that the Linkerd dashboard has been launched and that it's available on port 50750.

To view those metrics, you can use the Grafana dashboard, which is available at http://localhost:50750/grafana:

```
$ microk8s linkerd viz dashboard &
[1] 19577
$ Linkerd dashboard available at:
http://localhost:50750
Grafana dashboard available at:
http://localhost:50750/grafana
Opening Linkerd dashboard in the default browser
Failed to open Linkerd dashboard automatically
Visit http://localhost:50750 in your browser to view the dashboard
```

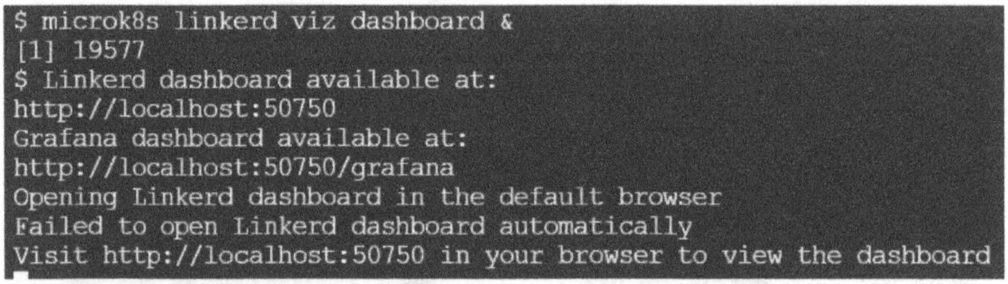

Figure 12.12 – Launching the Linkerd dashboard

The Linkerd dashboard gives you a bird's-eye view of what's going on with the services in real time. It can be used to see metrics such as the success rate, requests per second, and latency, as well as visualize service dependencies and understand the health of certain service routes:

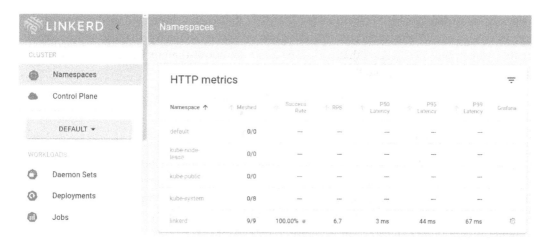

Figure 12.13 – The Linkerd dashboard

While the Linkerd dashboard gives you a bird's-eye view of what's going on with the services in real-time, Grafana dashboards, which are also part of the Linkerd control plane, provide usable dashboards for the services out of the box. These can also be used to monitor the services. Even for pods, we can get high-level stats and dive into the details:

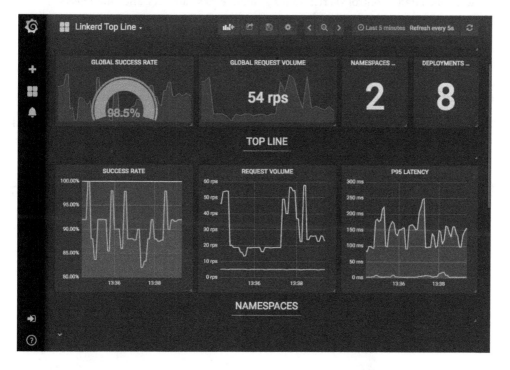

Figure 12. 14 – Linkerd Top Line metrics dashboard

To summarize, we have enabled Linkerd on the MicroK8s Kubernetes cluster and used it to monitor the services of a sample Nginx application. We also gathered relevant telemetry data such as the success rate, throughput, and latency. After that, we looked into a few out-of-the-box Grafana dashboards to see high-level metrics and dig into the details.

In the next section, we will look at Istio, another notable service mesh provider.

Overview of the Istio service mesh

Istio is an open source platform-independent service mesh that manages traffic, enforces policies, and collects telemetry. It is a platform for managing communication between microservices and applications. It also provides automated baseline traffic resilience, service metrics collection, distributed tracing, traffic encryption, protocol upgrades, and advanced routing functionality for all service-to-service communication without requiring changes to the underlying services.

The following are some of the vital features of Istio:

- Secure *service-to-service* communication via TLS encryption, strong identity-based authentication, and authorization

- Automatic load balancing for HTTP, gRPC, WebSocket, and TCP traffic

- Fine-grained traffic control via rich routing rules, retries, failovers, and fault injection

- A pluggable policy layer and configuration API that supports access controls, rate limits, and quotas

- Automatic metrics, logs, and traces for all cluster traffic, including cluster ingress and egress

An Istio service mesh is logically divided into two planes: a data plane and a control plane.

The data plane is made up of a collection of intelligent proxies that are deployed as sidecars. All network communication between microservices is mediated and controlled by these proxies. In addition, they collect and report telemetry on all mesh traffic.

The control plane is in charge of managing and configuring the proxies that are used to route traffic.

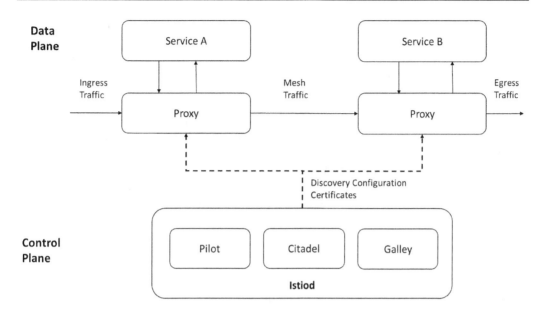

Figure 12.15 – Istio components

Now that we've provided a high-level overview and looked at the architecture, let's look at each component in more detail:

- **Istiod**: This manages service discovery, configuration, and certificates. It translates high-level routing rules that govern traffic behavior into Envoy-specific configurations and propagates them to sidecars at runtime. The Pilot component abstracts platform-specific service discovery mechanisms and synthesizes them into a standard format that can be consumed by any Envoy API-compliant sidecar. Istio also supports discovery for a variety of environments, including Kubernetes and virtual machines.

- **Envoy**: This is a high-performance proxy that mediates all inbound and outbound traffic for all services in the service mesh. The only Istio components that interact with data plane traffic are envoy proxies. Envoy proxies are deployed as service sidecars, logically augmenting the services with Envoy's many built-in features, such as the following:

  - Dynamic service discovery

  - Load balancing

  - TLS termination

  - HTTP/2 and gRPC proxies

  - Circuit breakers

  - Health checks

- Staged rollouts with a percentage-based traffic split

- Fault injection

- Rich metrics

The following are various components of the Istio system, as well as the abstractions that it employs:

- **Traffic management**: The traffic routing rules provided by Istio allow you to easily control the flow of traffic and API calls between services. Istio makes it simple to configure service-level properties such as circuit breakers, timeouts, and retries, as well as important tasks such as A/B testing, canary rollouts, and staged rollouts with percentage-based traffic splits. It also includes out-of-the-box reliability features that aid in making the application more resilient to failures of dependent services or the network.

- **Observability**: For all mesh service communications, Istio creates extensive telemetry. This telemetry lets users observe service behavior, allowing them to debug, maintain, and optimize their applications without putting additional strain on service developers. Users can acquire a comprehensive picture of how monitored services interact with one another and with Istio components.

Istio creates the following kinds of telemetry to give total service mesh observability:

- **Metrics**: Based on the four monitoring attributes, Istio creates a set of service metrics (latency, traffic, errors, and saturation). In addition, Istio provides extensive mesh control plane measurements. A basic set of mesh monitoring dashboards is supplied on top of these metrics.

- **Distributed tracing**: Dispersed traces result in distributed trace spans for each service, providing users with a comprehensive view of call flows and the service relationships inside a mesh.

- **Access logs**: As traffic flows into the service within a mesh, Istio generates a complete record of each request, including source and destination metadata. Users can utilize this information to examine service behavior down to individual workload instances.

- **Security**: To protect hosted services as well as data, Istio security includes strong identity, powerful policy management, transparent TLS encryption, authentication, and audit tools.

Now that we have covered the fundamentals, we can proceed to the next step, which is to enable the Istio add-on and run a sample application.

Enabling the Istio add-on and running a sample application

In this section, you will enable the Istio add-on in your MicroK8s Kubernetes cluster. Then, you will launch a sample application to show off Istio's capabilities.

> **Note**
>
> I'll be using an Ubuntu virtual machine for this section. The instructions for setting up a MicroK8s cluster are the same as those in *Chapter 5, Creating and Implementing Updates on Multi-Node Raspberry Pi Kubernetes Cluster*.

Step 1 – Enabling the Istio add-on

Use the following command to enable the Istio add-on:

```
microk8s enable istio
```

The following output indicates that the Istio add-on has been enabled:

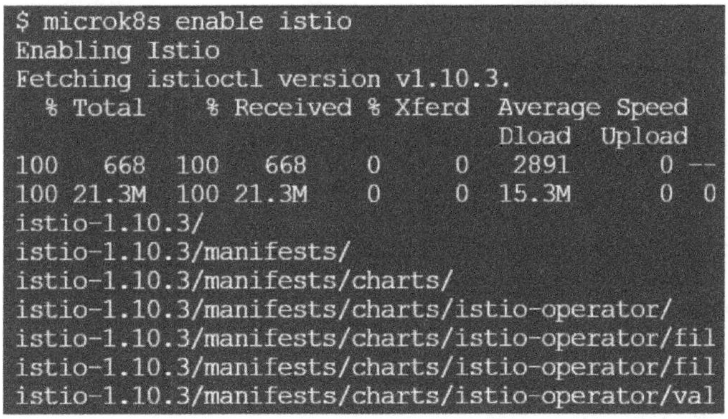

```
$ microk8s enable istio
Enabling Istio
Fetching istioctl version v1.10.3.
  % Total    % Received % Xferd  Average Speed
                                  Dload  Upload
100   668  100   668    0     0   2891      0 --
100 21.3M  100 21.3M    0     0   15.3M      0  0
istio-1.10.3/
istio-1.10.3/manifests/
istio-1.10.3/manifests/charts/
istio-1.10.3/manifests/charts/istio-operator/
istio-1.10.3/manifests/charts/istio-operator/fil
istio-1.10.3/manifests/charts/istio-operator/fil
istio-1.10.3/manifests/charts/istio-operator/val
```

Figure 12.16 – Enabling the Istio add-on

It will take some time to finish activating the add-on. The following output shows that Istio has been successfully enabled:

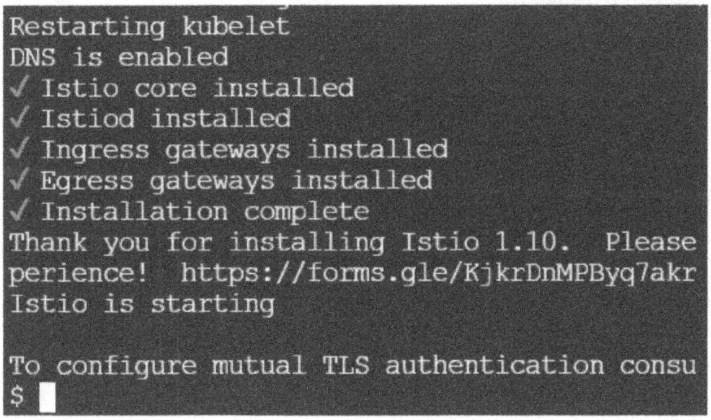

```
Restarting kubelet
DNS is enabled
✓ Istio core installed
✓ Istiod installed
✓ Ingress gateways installed
✓ Egress gateways installed
✓ Installation complete
Thank you for installing Istio 1.10.  Please
perience!  https://forms.gle/KjkrDnMPByq7akr
Istio is starting

To configure mutual TLS authentication consu
$ ▮
```

Figure 12.17 – Istio add-on enabled

Before we move on to the next step, let's make sure that all of the Istio components are up and running by using the following command:

```
kubectl get pods -n istio-system
```

The following output indicates that all the components are Running:

```
$ kubectl get pods -n istio-system
NAME                                        READY    STATUS
istiod-6f94fb9786-zfjn6                     1/1      Running
istio-ingressgateway-f9cd5d59d-kx9w8        1/1      Running
istio-egressgateway-77c5c9d46d-gfhk4        1/1      Running
$
```

Figure 12.18 – The Istio pods are running

Now that the Istio add-on has been enabled, let's deploy the sample application.

Step 2 – Deploying the sample application

Before deploying the sample Nginx application, we need to label the namespace as istio-injection=enabled so that Istio can inject sidecars into the deployment's pods.

Use the following command to label the namespace:

```
microk8s kubectl label namespace default istio-injection=enabled
```

The following output indicates that there is no error in the deployment. Now, we can deploy the sample application:

```
$ microk8s kubectl label namespace default istio-injection=enabled
namespace/default labeled
```

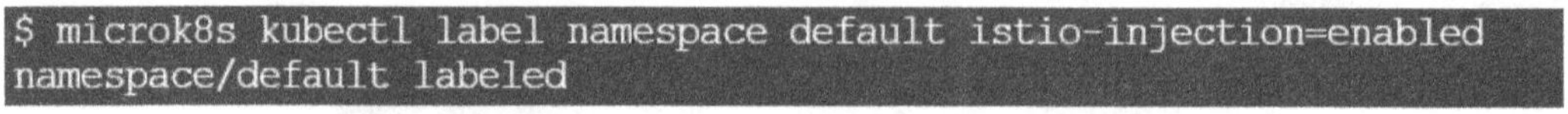

Figure 12.19 – Labeling the namespace

Use the following command to create a sample Nginx deployment:

```
kubectl apply -f https://k8s.io/examples/application/
deployment.yaml
```

The following output indicates that there is no error in the deployment. Now, we can ensure that Istio has been injected into pods:

```
$ kubectl apply -f https://k8s.io/examples/application/deployment.yaml
deployment.apps/nginx-deployment created
$
```

Figure 12.20 – Sample application deployment

With the deployment completed, we can check if the Istio labels have been added to the sample application deployment by using the `kubectl describe` command.

The following output confirms that the Istio labels have been added:

```
$ kubectl describe pod nginx-deployment-9456bbbf9-w2ltq
Name:          nginx-deployment-9456bbbf9-w2ltq
Namespace:     default
Priority:      0
Node:          host01/10.0.0.21
Start Time:    Sun, 17 Apr 2022 12:39:51 +0000
Labels:        app=nginx
               istio.io/rev=default
               pod-template-hash=9456bbbf9
               security.istio.io/tlsMode=istio
               service.istio.io/canonical-name=nginx
               service.istio.io/canonical-revision=latest
Annotations:   cni.projectcalico.org/podIP: 10.1.239.199/32
               cni.projectcalico.org/podIPs: 10.1.239.199/32
               kubectl.kubernetes.io/default-container: ngin
               kubectl.kubernetes.io/default-logs-container:
               prometheus.io/path: /stats/prometheus
               prometheus.io/port: 15020
               prometheus.io/scrape: true
               sidecar.istio.io/status:
                 {"initContainers":["istio-init"],"container
istio-data","istio-podinfo","istio-token","istiod-...
Status:        Running
```

Figure 12.21 – Istio annotations added

We can also use the `istioctl` CLI command to get an overview of the Istio mesh:

```
microk8s istioctl proxy-status
```

The following output indicates that our sample Nginx deployment has been SYNCED with the Istiod control plane:

```
$ microk8s istioctl proxy-status
NAME                                                        CDS       LDS       EDS       RDS
OD                       VERSION
istio-egressgateway-77c5c9d46d-qfhk4.istio-system           SYNCED    SYNCED    SYNCED    NOT SENT
od-6f94fb9786-zfjn6      1.10.3
istio-ingressgateway-f9cd5d59d-kx9w8.istio-system           SYNCED    SYNCED    SYNCED    NOT SENT
od-6f94fb9786-zfjn6      1.10.3
nginx-deployment-9456bbbf9-8z6dm.default                    SYNCED    SYNCED    SYNCED    SYNCED
od-6f94fb9786-zfjn6      1.10.3
nginx-deployment-9456bbbf9-w2ltq.default                    SYNCED    SYNCED    SYNCED    SYNCED
od-6f94fb9786-zfjn6      1.10.3
$
```

Figure 12.22 – Istio proxy status

If any of the sidecars aren't receiving configuration or are out of sync, then you can use the `proxy-status` command.

If a proxy isn't listed, it's because it's not currently linked to an Istiod instance:

- `SYNCED` indicates that the Envoy proxy has acknowledged the most recent configuration that's been supplied to it by Istiod.

- `NOT SENT` indicates that Istiod has not sent any messages to the Envoy proxy. This is frequently because Istiod has nothing to send.

- `STALE` indicates that Istiod sent an update to the Envoy proxy but did not receive a response. This usually indicates a problem with networking between the Envoy proxy and Istiod, or a flaw with Istio itself.

Congratulations! You have added Istio proxies to the sample application! We added Istio to existing services without touching the original YAML.

Step 3 – Exploring the Istio service dashboard

For all service communication within a mesh, Istio creates extensive telemetry. This telemetry allows service behavior to be observed, allowing service mesh users to troubleshoot, maintain, and optimize their applications without adding to the workload for service developers.

As we discussed previously, to enable overall service mesh observability, Istio creates the following forms of telemetry:

- **Metrics**: Based on monitoring performance, Istio generates a set of service metrics (latency, traffic, errors, and saturation). For the mesh control plane, Istio also provides detailed metrics. On top of these metrics, a default set of mesh monitoring dashboards is given:

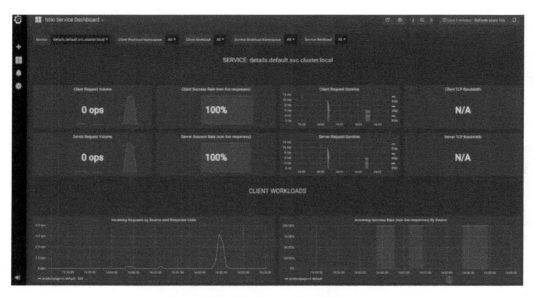

Figure 12.23 – The Istio service dashboard

The resource usage dashboard looks as follows. This is where we can get details about memory, CPU, and disk usage:

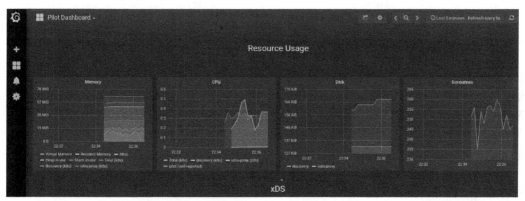

Figure 12.24 – Istio Resource Usage dashboard

- **Distributed traces**: Istio creates distributed trace spans for each service, giving users a complete picture of the call flows and service dependencies in a mesh:

Figure 12.25 – Istio distributed traces

- **Access logs**: Istio generates a full record of each request as traffic flows into a service within a mesh, including source and destination metadata. Users can audit service behavior down to the individual workload instance level using this data.

To summarize, we deployed Istio on the MicroK8s Kubernetes cluster and used it to monitor a sample Nginx application's services. We also had a look at the Istio service dashboard, which allows us to examine telemetry data to debug, maintain, and improve applications. Finally, we looked at how metrics, distributed traces, and access logs can be used to enable overall service mesh observability.

In short, a service mesh provides uniform discovery, security, tracing, monitoring, and failure management. So, if a Kubernetes cluster has a service mesh, you can have the following without changing the application code:

- Automatic load balancing
- Fine-grained control of traffic behavior with routing rules, retries, failovers, and more
- Pluggable policy layer
- A configuration API that supports access control, rate limits, and quotas
- Service discovery
- Service monitoring with automatic metrics, logs, and traces for all traffic
- Secure service-to-service communication

In most implementations, the service mesh serves as the single pane of glass for a microservices architecture. It's where you go to troubleshoot problems, enforce traffic policies, set rate limits, and test new code. It serves as your central point for monitoring, tracing, and controlling the interactions of all services – that is, how they are connected, performed, and secured. In the next section, we will look at some of the most common use cases.

Common use cases for a service mesh

A service mesh is useful for any type of microservices architecture from an operations standpoint. This is because it allows you to control traffic, security, permissions, and observability. Here are some of the most common, standardized, and widely accepted use cases for service meshes today:

- **Improving the observability**: Through service-level visibility, tracing, and monitoring, we may improve the observability of distributed services. Some of the service mesh's primary features boost visibility and your ability to troubleshoot and manage situations dramatically. For example, if one of the architecture's services becomes a bottleneck, retrying is a frequent option, although this may exacerbate the bottleneck due to timeouts. With a service mesh, you can quickly break the circuit to failing services, disable non-functioning replicas, and maintain the API's responsiveness.

- **Blue/Green deployments**: A service mesh allows you to leverage Blue/Green deployments to successfully roll out new application upgrades without them affecting services due to its traffic control features. You begin by exposing the new version to a limited group of users, testing it, and then rolling it out to all production instances.

- **Chaos monkey/production testing**: To improve deployment robustness, the ability to inject delays and errors is also available.

- **Modernizing your legacy applications**: You can utilize a service mesh as an enabler while decomposing your apps if you're in the process of upgrading your old applications to Kubernetes-based microservices. You can register your existing applications as services in the service catalog and then migrate them to Kubernetes over time without changing the communication style between them.

- **The API Gateway technique**: With the help of a service mesh, you may leverage the API Gateway technique for service-to-service connectivity and complicated API management schemes within your clusters. A service mesh acts as superglue, dynamically connecting microservices with traffic controls, restrictions, and testing capabilities.

Many new and widely accepted use cases will join those listed previously as service meshes become more popular. Now, let's look at the considerations for choosing a service mesh provider.

Guidelines on choosing a service mesh

In this section, we will provide a brief comparison of the features offered by service mesh providers. Choosing one that meets your fundamental requirements boils down to whether or not you want more than just the essentials. Istio provides the most features and versatility, but keep in mind that flexibility equals complexity. Linkerd may be the best option for a basic strategy that only supports Kubernetes:

Parameters	Istio	Linkerd
Supported Workloads	Kubernetes and VMs	Kubernetes
Single Point of Failure	No	No
Sidecar Proxy	Yes – Envoy	Yes
Support for mTLS	Yes	Yes
Support for Certificate Management	Yes	Yes
Support for Authentication and Authorization	Yes	Yes
Support for the TCP Protocol	Yes	Yes
Support for the HTTP/1.x/2 Protocol	Yes	Yes
Support for the gRPC Protocol	Yes	Yes
Blue/Green Deployments Support	Yes	Yes
Circuit Breaking Support	Yes	No
Fault Injection Support	Yes	Yes
Rate Limiting Support	Yes	No
Chaos Monkey Style Testing Support	Yes	Limited
Monitoring (Prometheus Support)	Yes	Yes
Distributed Tracing Support	Yes	Limited
Multi-Cluster Support	Yes	No
Operations Complexity	High	Low
Native GUI	No	Yes

Table 12.1 – Comparison between the Istio and Linkerd service meshes

Now that we've seen some recommendations for selecting a service mesh, let's look at the best practices for configuring one.

Best practices for configuring a service mesh

Although a service mesh is extremely beneficial to development teams, putting one in place requires some effort. A service mesh gives you a lot of flexibility and room to tailor it to your needs because it has so many moving pieces. Flexibility usually comes at the expense of complexity. While working with a service mesh, the following best practices will provide you with some useful guidelines:

- **Adopt a GitOps approach**: Traffic regulations, rate limits, and networking setup are all part of the service mesh's configuration. The configuration can be used to install the service mesh from the ground up, update its versions, and migrate between clusters. As a result, it is recommended that the configuration be regarded as code and that GitOps be utilized in conjunction with a continuous deployment pipeline.

- **Use fewer clusters**: Fewer clusters with a big number of servers perform better than many clusters with fewer instances for service mesh products. As a result, it's best to keep the number of redundant clusters as low as possible, allowing you to take advantage of your service mesh approach's straightforward operation and centralized configuration.

- **Use appropriate monitoring alerts and request tracing**: Service mesh apps are advanced applications that manage the traffic of increasingly complicated distributed applications. For system observability, metric collection, visualization, and dashboards are essential. Using Prometheus or Grafana, which will be offered by your service mesh, you can create alerts according to your requirements.

- **Focus on comprehensive security**: The majority of service mesh systems offer mutual TLS, certificate management, authentication, and authorization as security features. To limit communication across clustered apps, you can design and enforce network policies. However, it should be emphasized that designing network policies is not a simple operation. You must consider future scalability and cover all eventualities for currently running apps. As a result, using a service mesh to enforce network policies is inconvenient and prone to errors and security breaches. Who is transmitting or receiving data is unimportant to service mesh solutions. Any hostile or malfunctioning application can retrieve your sensitive data if network policies allow it. As a result, rather than depending exclusively on the security features of service mesh devices, it's essential to think about the big picture.

In conclusion, a service mesh allows you to decouple the application's business logic from observability, network, and security policies. You can connect to, secure, and monitor your microservices with it.

Summary

The number of services that make up an application grows dramatically as monolithic apps are split down into microservices. And managing a huge number of entities isn't easy. By standardizing and automating communication between services, a Kubernetes native service mesh, such as Istio or Linkerd, tackles difficulties created by container and service sprawl in a microservices architecture. Security, service discovery, traffic routing, load balancing, service failure recovery, and observability are all standardized and automated by a service mesh.

In this chapter, we learned how to enable the Linkerd and Istio add-ons and inject sidecars into sample applications. Then, we examined the respective dashboards, which allowed us to examine telemetry data to debug, maintain, and improve applications. We also examined how metrics, distributed traces, and access logs can be used to improve overall service mesh observability.

After that, we looked at some of the most prevalent use cases for service meshes today, as well as some tips on how to pick the right service mesh. We also provided a list of service mesh configuration best practices.

In the next chapter, you will learn how to set up a highly available cluster. A highly available Kubernetes cluster can withstand a component failure and continue to serve workloads without interruption.

13

Resisting Component Failure Using HA Clusters

In the previous chapter, we looked at how to enable Linkerd or Istio service mesh add-ons and inject sidecars into a sample application. We also looked at the dashboards that allow us to look at telemetry data in order to troubleshoot, manage, and improve applications. We then looked at how metrics, distributed traces, and access logs can help with overall service mesh observability. We additionally looked at some of the most common service mesh use cases today, as well as some recommendations for how to choose the correct service mesh. We also covered a list of service mesh configuration best practices.

Through the use of dynamic container scheduling, Kubernetes offers higher reliability and resiliency for distributed applications. But how can you ensure that Kubernetes itself remains operational when a component, or even an entire data center site, fails? In this chapter, we will look into our next use case on how to configure Kubernetes for a **high-availability (HA)** cluster.

HA keeps applications up and running even if the site fails partially or completely. The basic goal of HA is to eliminate potential points of failure. It can be achieved at many levels of infrastructure and within different cluster components. However, the amount of availability that fits a particular situation is determined by a number of factors, including your business needs, service-level agreements with your customers, and resource availability.

Kubernetes aims to provide HA for both applications and infrastructure. Each of the control plane (master) components can be configured for multi-node replication (multi-master setup) to improve availability. However, it's important to remember that HA and a multi-master setup are not synonymous. Even if you have three or more control plane nodes and only one NGINX instance front-load balancing to those masters, you have a multi-master cluster setup, but not an HA setup, because NGINX could still go down at any time and cause failures.

Control plane nodes are vital because they operate the services that control, monitor, and maintain the Kubernetes cluster's state. The API server, cluster state storage, the scheduler, and the controller manager are all part of the control plane. If only one control plane node fails in a cluster, the cluster's operation and stability may be seriously impaired. HA clusters address this by running numerous control plane nodes at the same time, and while this doesn't completely eliminate risk, it reduces it greatly.

MicroK8s' HA option has been simplified and enabled by default. This means that a cluster can survive a node failure and continue to serve workloads without interruption. HA is a critical feature for enterprises wishing to deploy containers and pods that can provide the level of stability required when working at scale.

Canonical's lightweight Dqlite SQL database is used to enable the HA clustering functionality. By embedding the database into Kubernetes, Dqlite reduces the cluster's memory footprint and eliminates process overhead. This is significant for IoT and Edge applications. The deployment of resilient Kubernetes clusters at the edge is simplified when Dqlite is used as the Kubernetes datastore. Edge applications can now achieve exceptional reliability at a low cost on x86 or ARM commodity appliances such as clusters of Intel NUCs or Raspberry Pi boards. In this chapter, we're going to cover the following main topics:

- An overview of HA topologies

- Setting up an HA Kubernetes cluster

- Kubernetes HA best practices

An overview of HA topologies

In this section, we'll look at the two most common HA topologies for enabling HA clusters. The Kubernetes cluster's control plane is mostly stateless. The cluster datastore, which operates as the one source of truth for the whole cluster, is the only stateful component of the control plane. Internal and external consumers can access and alter the state through the API server, which serves as a gateway to the cluster datastore. MicroK8s uses Dqlite, a distributed and highly accessible variant of SQLite, as the key-value database to preserve the cluster's state.

Before looking at the HA topologies, let us look at potential failure scenarios that could hamper cluster operations:

- **Loss of control plane (master) node**: Loss of the master node or its services will have a major impact. The cluster will be unable to respond to API commands or the deployment of nodes. Each service in the master node, as well as the storage layer, is crucial and must be designed for HA.

- **Loss of cluster datastore**: Whether the cluster datastore is run on the master node or set up separately, losing cluster data is disastrous because it contains all cluster information. To avoid this, the cluster datastore must be configured in an HA cluster.

- **Worker node(s) failure**: In most circumstances, Kubernetes will be able to identify and failover pods automatically. The end users of the application may not notice any difference, depending on how the services are load-balanced. If any pods on a node become unresponsive, kubelet will detect this and notify the master to start another pod.
- **Network failures**: Network outages and partitions can cause the master and worker nodes in a Kubernetes cluster to become unreachable. In some circumstances, they will be classified as node failures.

Now that we've seen some of the probable failure situations, let's look at how they can be mitigated using HA topologies that can withstand the failure of one or more master nodes while running Kubernetes production workloads.

The topology of an HA Kubernetes cluster can be configured in two ways, depending on how the cluster datastore is configured. The first topology is based on a stacked cluster design, in which each node hosts a Dqlite instance, as well as the control plane. The `kube-apiserver` instance, the `kube-scheduler` instance, and the `kube-controller-manager` instance are running on each control plane node. A load balancer exposes the `kube-apiserver` instance to worker nodes.

Each control plane node produces a local Dqlite member, which exclusively communicates with this node's `kube-apiserver` instance. The local `kube-controller-manager` instance and the `kube-scheduler` instance are the same.

The control planes and local Dqlite members are linked on the same node in this topology. It is easier to set up and administer for replication. However, a stacked cluster is vulnerable to failed coupling. When one node fails, the local Dqlite members and a control plane instance are lost, putting redundancy at risk. This threat can be reduced by adding more control plane nodes.

Hence for an HA Kubernetes cluster, this design necessitates at least three stacked control plane nodes, as depicted in *Figure 13.1*:

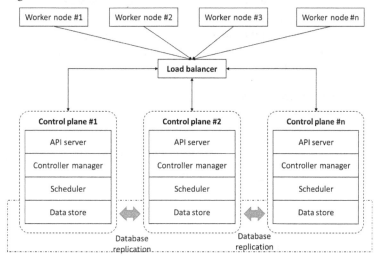

Figure 13.1 – A stacked control plane topology

The second topology makes use of an external Dqlite cluster that is installed and controlled on a separate set of hosts.

Each control plane node in this architecture runs a `kube-apiserver` instance, a `kube-scheduler` instance, and a `kube-controller-manager` instance, with each Dqlite host communicating with the `kube-apiserver` instance of each control plane node:

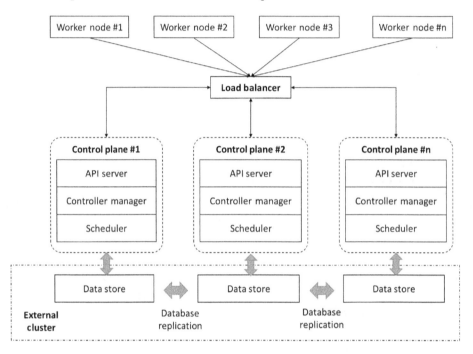

Figure 13.2 – The topology of an external cluster datastore

The control plane and local Dqlite member are decoupled in this topology. As a result, it provides an HA configuration in which losing a control plane instance or Dqlite member has less of an impact and does not influence cluster redundancy as much as the stacked HA architecture.

This design, however, requires twice as many hosts as the stacked HA topology. An HA cluster with this topology requires a minimum of three hosts for control plane nodes and three hosts for Dqlite nodes.

In the next section, we are going to walk through the steps involved in setting up an HA cluster using the stacked cluster HA topology. All that is necessary for HA MicroK8s is three or more nodes in the cluster, after which Dqlite becomes highly available automatically. If the cluster has more than three nodes, additional nodes will be designated as standby candidates for the datastore and promoted automatically if one of the data store's nodes fails. The automatic promotion of standby nodes into the Dqlite voting cluster makes HA MicroK8s self-sufficient and ensures that a quorum is maintained even if no administrative action is performed.

Setting up an HA Kubernetes cluster

We are going to configure and implement an HA MicroK8s Kubernetes cluster utilizing the stacked cluster HA topology that we discussed before. We'll use the three nodes to install and configure MicroK8s on each of the nodes and simulate node failure to see whether the cluster is resisting component failures and functioning as expected.

To recap, a control plane is run by all the nodes in the HA cluster. A portion (at least three) of the cluster nodes keeps a copy of the Kubernetes cluster datastore (the Dqlite database). A voting procedure is used to pick a leader for database maintenance. Aside from the voting nodes, there are non-voting nodes that store a copy of the database discreetly. These nodes are ready to replace a leaving voter. Finally, some nodes do not vote or duplicate the database. These are known as spare nodes. To summarize, the three node roles are as follows:

- **Voters**: Replicating the database, participating in leader election
- **Standby**: Replicating the database, *not* participating in leader election
- **Spare**: *Not* replicating the database, *not* participating in leader election

The administrator doesn't need to monitor how cluster formation, database syncing, or voter and leader elections are all done since it's transparent and taken care of. *Figure 13.3* depicts our Raspberry Pi cluster setup:

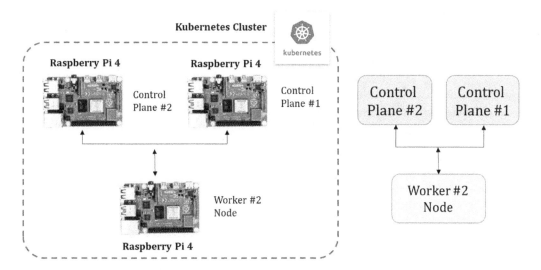

Figure 13.3 – A fully functional HA cluster setup

Now that we know what we want to do, let's look at the requirements.

Requirements

Before you begin, the following are the prerequisites for building a Raspberry Pi Kubernetes cluster:

- A microSD card (4 GB minimum, 8 GB recommended)
- A computer with a microSD card drive
- A Raspberry Pi 2, 3, or 4 (with three nodes)
- A micro-USB power cable (a USB-C cable for the Pi 4)
- A Wi-Fi network or an Ethernet cable with an internet connection
- A monitor with an HDMI interface (optional)
- An HDMI cable for the Pi 2 and 3 and a micro-HDMI cable for the Pi 4 (optional)
- A USB keyboard (optional)

Now that we've established what the requirements are for setting up an HA MicroK8s Kubernetes cluster, we'll move on to the step-by-step instructions on how to complete it.

Step 1 – Creating the MicroK8s Raspberry Pi cluster

Please follow the steps that we covered in *Chapter 5*, *Creating and Implementing Updates on Multi-Node Raspberry Pi Kubernetes Clusters*, to create the MicroK8s Raspberry Pi cluster. Here's a quick refresher:

1. Install an OS image to the SD card:

   A. Configure Wi-Fi access settings.

   B. Configure remote access settings.

   C. Configure control group settings.

   D. Configure a hostname.

2. Install and configure MicroK8s.

3. Add additional control plane nodes and worker nodes to the cluster.

> **Note**
>
> Starting from the MicroK8s 1.23 release, there is now an option to add worker-only nodes. This type of node does not execute the control plane and does not contribute to the cluster's HA. They, on the other hand, utilize fewer resources and are hence appropriate for low-end devices. Worker-only nodes are also appropriate in systems where the nodes executing the Kubernetes workloads are unreliable or cannot be trusted to hold the control plane.
>
> To add a worker-only node to the cluster, use the `--worker` flag when running the `microk8s join` command:
>
> ```
> microk8s join 192.168.1.8:25000/92b2db237428470dc4fcfc4ebbd9dc81/
> 2c0cb3284b05 --worker
> ```
>
> A Traefik load balancer runs on a worker node, allowing communication between local services (kubelet and kube-proxy) and API servers operating on several control plane nodes. When adding a worker node, MicroK8s tries to discover all API server endpoints in the cluster and correctly set up the new node. We will not use worker-only nodes in this section, but worker nodes that also host the control plane instead.

We'll repeat the methods set out in *Chapter 5, Creating and Implementing Updates on Multi-Node Raspberry Pi Kubernetes Clusters*, for our present setup, as shown in the following table:

Hostname	IP address	Role
Controlplane1	192.168.1.6	Master
Controlplane2	192.168.1.7	Master
Worker2	192.168.1.8	Master, Worker

Table 13.1 – A Raspberry Pi cluster setup

Now that we're clear on our goals, we'll take installing and configuring MicroK8s on each Raspberry Pi board step by step and then combine multiple deployments to build a fully functional cluster.

Installing and configuring MicroK8s

SSH into your control plane node and install the MicroK8s snap:

```
sudo snap install microk8s --classic
```

The following command execution output confirms that the MicroK8s snap has been installed successfully:

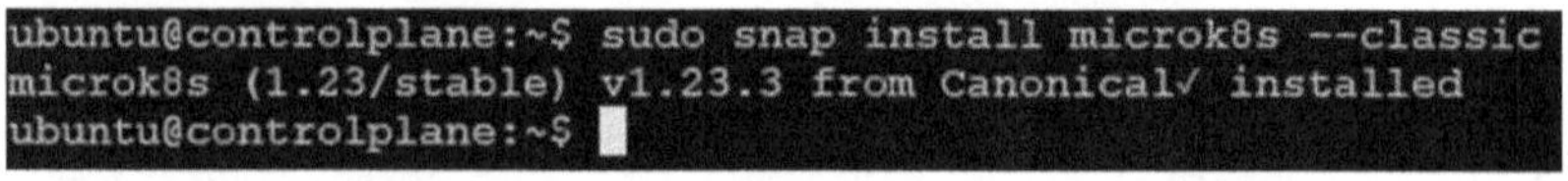

```
ubuntu@controlplane:~$ sudo snap install microk8s --classic
microk8s (1.23/stable) v1.23.3 from Canonical√ installed
ubuntu@controlplane:~$ 
```

Figure 13.4 – MicroK8s installation

The following command execution output confirms that MicroK8s is running successfully:

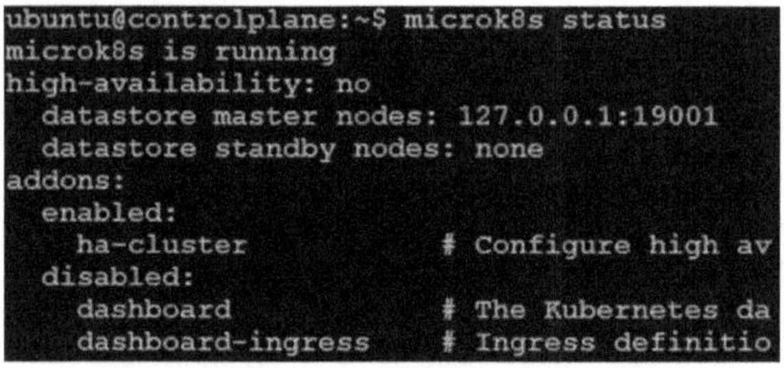

```
ubuntu@controlplane:~$ microk8s status
microk8s is running
high-availability: no
  datastore master nodes: 127.0.0.1:19001
  datastore standby nodes: none
addons:
  enabled:
    ha-cluster              # Configure high av
  disabled:
    dashboard               # The Kubernetes da
    dashboard-ingress       # Ingress definitio
```

Figure 13.5 – Inspect your MicroK8s cluster

If the installation is successful, then you should see the following output:

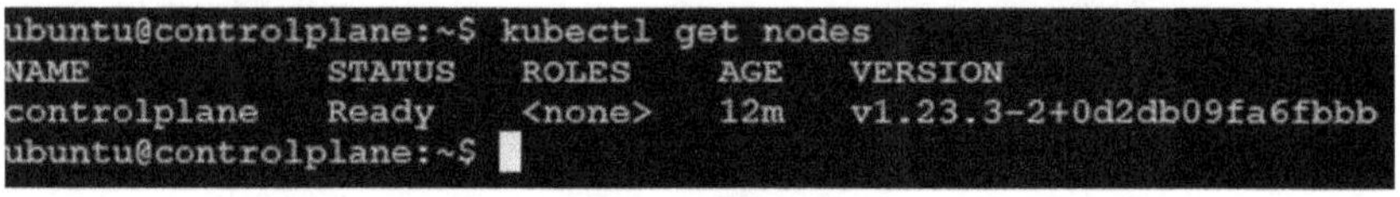

```
ubuntu@controlplane:~$ kubectl get nodes
NAME           STATUS    ROLES     AGE     VERSION
controlplane   Ready     <none>    12m     v1.23.3-2+0d2db09fa6fbbb
ubuntu@controlplane:~$ 
```

Figure 13.6 – Verify whether the node is in a ready state

Repeat the MicroK8s installation process on the other nodes as well.

The next step is to add a control plane node and worker node to the cluster. Open PuTTY shell to control plane node and run the following command to generate the connection string:

```
sudo microk8s.add-node
```

The following command execution output validates that the command was successfully executed and provides instructions for the connection string:

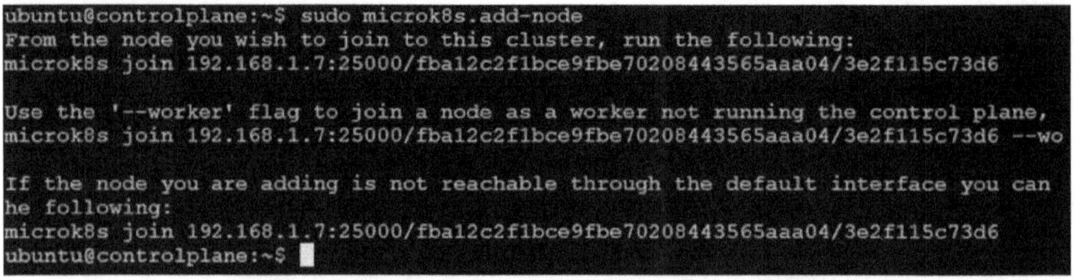

```
ubuntu@controlplane:~$ sudo microk8s.add-node
From the node you wish to join to this cluster, run the following:
microk8s join 192.168.1.7:25000/fba12c2f1bce9fbe70208443565aaa04/3e2f115c73d6

Use the '--worker' flag to join a node as a worker not running the control plane,
microk8s join 192.168.1.7:25000/fba12c2f1bce9fbe70208443565aaa04/3e2f115c73d6 --wo

If the node you are adding is not reachable through the default interface you can
he following:
microk8s join 192.168.1.7:25000/fba12c2f1bce9fbe70208443565aaa04/3e2f115c73d6
ubuntu@controlplane:~$ 
```

Figure 13.7 – Generate connection string for adding nodes

As indicated by the preceding command execution output, the connection string is generated in the form of `<control plane_ip>:<port>/<token>`.

Adding additional control plane nodes

We now have the connection string to join with the control plane node. Open the PuTTY shell to the `controlplane1` node and run the `join` command to add it to the cluster:

```
microk8s join <control plane_ip>:<port>/<token>
```

The command was successfully executed, and the node has joined the cluster, as shown in the output:

```
ubuntu@controlplane1:~$ sudo microk8s join 192.168.1.7:25000/fba12c2f1bce9fbe
65aaa04/3e2f115c73d6
Contacting cluster at 192.168.1.7

The node has joined the cluster and will appear in the nodes list in a few se
ds.

Currently this worker node is configured with the following kubernetes API se
r endpoints:
    - 192.168.1.7 and port 16443, this is the cluster node contacted during
join operation.

If the above endpoints are incorrect, incomplete or if the API servers are be
d a loadbalancer please update
/var/snap/microk8s/current/args/traefik/provider.yaml
```

Figure 13.8 – Adding an additional control plane node to the cluster

Our next step is to add a worker node to the cluster now that we have added an additional control plane node so that we can simulate node failure to see whether the cluster is able to resist component failures and function as we expect.

Adding a worker node

We now have the connection string to join with the control plane node. Open the PuTTY shell to the worker node and run the `join` command to add it to the cluster:

```
microk8s join <control plane_ip>:<port>/<token>
```

The command was successfully executed, and the node has joined the cluster, as shown in the output:

```
ubuntu@worker2:~$ sudo microk8s join 192.168.1.7:25000/fba12c2f1bce9fbe702084435
65aaa04/3e2f115c73d6
Contacting cluster at 192.168.1.7

The node has joined the cluster and will appear in the nodes list in a few secon
ds.

Currently this worker node is configured with the following kubernetes API serve
r endpoints:
    - 192.168.1.7 and port 16443, this is the cluster node contacted during the
join operation.

If the above endpoints are incorrect, incomplete or if the API servers are behin
d a loadbalancer please update
/var/snap/microk8s/current/args/traefik/provider.yaml
```

Figure 13.9 – Adding a worker node to the cluster

As indicated by the preceding command execution output, you should be able to see the new node in a few seconds on the control plane.

Use the following command to verify whether the new node has been added to the cluster:

```
kubectl get nodes
```

The following command execution output shows that `controlplane`, `controlplane1`, and `worker2` are part of the cluster:

```
ubuntu@controlplane:~$ microk8s kubectl get no
NAME            STATUS   ROLES    AGE      VERSION
controlplane    Ready    <none>   5h12m    v1.24.0-2+f76e51e86eadea
controlplane1   Ready    <none>   4h23m    v1.24.0-2+f76e51e86eadea
worker2         Ready    <none>   13m      v1.24.0-2+59bbb3530b6769
ubuntu@controlplane:~$
```

Figure 13.10 – The cluster is ready, and control planes and worker2 are part of the cluster

A fully functional multi-node Kubernetes cluster would look like that shown in *Figure 13.3*. To summarize, we have installed MicroK8s on the Raspberry Pi boards and joined multiple deployments to form the cluster. We've also added control plane nodes and worker nodes to the cluster.

Now that we have a fully functional cluster, we will move on to the next step of examining the HA setup.

Step 2 – Examining the HA setup

Since we have more than one node running the control plane, MicroK8s' HA would be achieved automatically. An HA Kubernetes cluster requires three conditions to be satisfied:

- At any given time, there must be more than one node available.

- The control plane must run on more than one node so that the cluster does not become unusable, even if a single node fails.

- The cluster state must be stored in a highly accessible datastore.

We can check the current state of the HA cluster using the following command:

```
microk8s status
```

The following command execution output confirms that HA has been achieved and also displays the datastore master nodes. Standby nodes are set to none since we have only three nodes; additional nodes will be designated as standby candidates for the datastore and will be promoted automatically if one of the datastore's nodes fails:

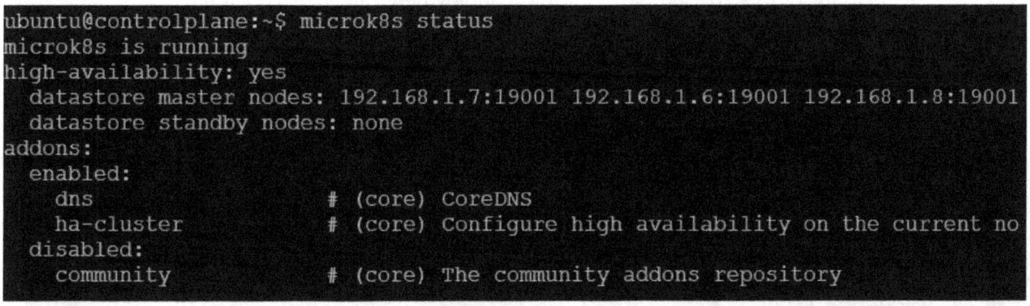

```
ubuntu@controlplane:~$ microk8s status
microk8s is running
high-availability: yes
  datastore master nodes: 192.168.1.7:19001 192.168.1.6:19001 192.168.1.8:19001
  datastore standby nodes: none
addons:
  enabled:
    dns                    # (core) CoreDNS
    ha-cluster             # (core) Configure high availability on the current no
  disabled:
    community              # (core) The community addons repository
```

Figure 13.11 – Inspecting the MicroK8s HA cluster

Congrats! You now have a secure, distributed, highly available Kubernetes cluster that's ready for a production-grade MicroK8s cluster environment. In the next section, we are going to deploy a sample application on the MicroK8s cluster that we just created.

Step 3 – Deploying a sample containerized application

In this section, we will deploy the NGINX deployment from the Kubernetes examples repository in our multi-node HA MicroK8s cluster setup.

The following command will deploy the sample application deployment:

```
kubectl apply -f https://k8s.io/examples/controllers/nginx-
deployment.yaml
```

The following command execution output indicates that there is no error in the deployment and in the next steps, we can verify this using the `get deployments` command:

```
ubuntu@controlplane:~$ kubectl apply -f https://k8s.io/examples/controllers/nginx-deployment.yaml
deployment.apps/nginx-deployment created
ubuntu@controlplane:~$
```

Figure 13.12 – A sample application deployment

The following command execution output displays the information about the deployment:

```
ubuntu@controlplane:~$ kubectl get deployments
NAME                 READY   UP-TO-DATE   AVAILABLE   AGE
nginx-deployment     3/3     3            3           4m7s
ubuntu@controlplane:~$
```

Figure 13.13 – The sample application deployments are in ready state

Let us also check where the pods are running using the following command:

```
kubectl get pod -o=custom-columns=NODE:.spec.nodeName,NAME:.
metadata.name
```

The following command execution output indicates that pods are equally distributed between the nodes:

```
ubuntu@controlplane:~$ kubectl get pod -o=custom-columns=NODE:.spec.nodeName,NAME:.metadata.name
NODE            NAME
controlplane    nginx-deployment-6595874d85-k44jr
controlplane1   nginx-deployment-6595874d85-d9zbd
worker2         nginx-deployment-6595874d85-kql6z
ubuntu@controlplane:~$
```

Figure 13.14 – The pod distribution across the nodes

Great! We have just deployed our sample application on the Raspberry multi-node HA cluster. To summarize, we built a Kubernetes Raspberry Pi cluster and used it to deploy a sample application. We'll perform some of the tests to check whether our cluster is resistant to failures in the next step.

> **Note**
> In the case that add-ons are enabled on a multi-node HA cluster, if the client binaries are downloaded and installed for an add-on, those binaries will only be available on the specific node from which the add-on was enabled.

Step 3 – Simulating control plane node failure

To simulate node failure, we will use the `cordon` command to mark the node as `unschedulable`. If the node is unschedulable, the Kubernetes controller will not schedule new pods on this node.

Let's use `cordon` on the `controlplane1` node so that we can simulate control plane failure. Use the following command to cordon the node:

```
kubectl cordon controlplane1
```

The following command execution output shows that `controlplane1` has been cordoned:

```
ubuntu@controlplane:~$ kubectl cordon controlplane1
node/controlplane1 cordoned
ubuntu@controlplane:~$
```

Figure 13.15 – Cordoning the controlplane1 node

Even though the `controplane1` node is cordoned, existing pods will still run. We can now use the `drain` command to delete all the pods. Use the following command to drain the node:

```
kubectl drain –force --ignore-daemonsets controlplane1
```

Use the `--ignore-daemonsets` flag to drain the nodes that contain the pods that are managed by DaemonSet.

The following command execution output shows that the pods running on `controlplane1` have been deleted successfully:

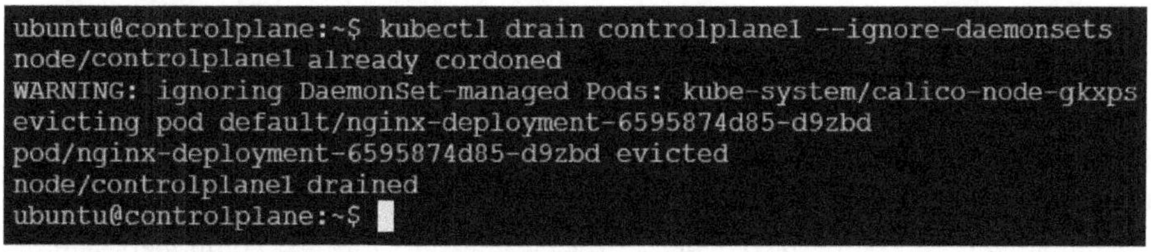

```
ubuntu@controlplane:~$ kubectl drain controlplane1 --ignore-daemonsets
node/controlplane1 already cordoned
WARNING: ignoring DaemonSet-managed Pods: kube-system/calico-node-gkxps
evicting pod default/nginx-deployment-6595874d85-d9zbd
pod/nginx-deployment-6595874d85-d9zbd evicted
node/controlplane1 drained
ubuntu@controlplane:~$
```

Figure 13.16 – Draining the controlplane1 node

Because the control plane is run by all nodes in the HA cluster, if one of the control plane nodes (`controlplane1`, for example) fails, cluster decisions can switch over to another control plane node and continue working without much disruption.

As part of the new control plane decisions, the Kubernetes controller will now recreate a new pod and schedule it in a different node as soon as the pod is deleted. It cannot be placed on the same node because the scheduling is disabled (since we have cordoned the `controlplane1` node).

Let us inspect where the pods are running using the `kubectl get pods` command. The following command execution output shows that the new pod has been rescheduled to the `controlplane` node:

```
ubuntu@controlplane:~$ kubectl get pod -o=custom-columns=NODE:.spec.nodeName,NAME:.metadata.name
NODE            NAME
controlplane    nginx-deployment-6595874d85-k44jr
worker2         nginx-deployment-6595874d85-kql6z
controlplane    nginx-deployment-6595874d85-kptt2
ubuntu@controlplane:~$
```

Figure 13.17 – Pod redistribution across the nodes

The following command execution output shows that the deployments have also been restored, despite the failure of one of the control plane nodes:

```
ubuntu@controlplane:~$ kubectl get deployments
NAME                READY   UP-TO-DATE   AVAILABLE   AGE
nginx-deployment    3/3     3            3           8m55s
ubuntu@controlplane:~$
```

Figure 13.18 – The deployments are in a ready state

Almost all HA cluster administration is invisible to the administrator and requires minimum configuration. Only the administrator has the ability to add and remove nodes. The following parameters should be considered to ensure the cluster's health:

- If the leader node is *removed*, such as by crashing and never returning, the cluster may take up to 5 seconds to elect a new leader.

- It can take up to 30 seconds to convert a non-voter into a voter. This promotion occurs when a new node joins the cluster or when a voter fails.

To summarize, we used the stacked cluster HA topology to configure and implement a highly available MicroK8s Kubernetes cluster. We used the three nodes to install and configure MicroK8s on each one, as well as simulating node failure to see whether the cluster can withstand component failures and continue to function as expected. In the next section, we will touch upon some of the best practices for implementing Kubernetes for a production-grade Kubernetes cluster.

Kubernetes HA best practices

As people become more acquainted with Kubernetes, there are trends toward more advanced use of the platform, such as users deploying Kubernetes in an HA architecture to ensure full production uptime. According to the recent *Kubernetes and cloud native operations report, 2022* (https://juju.is/cloud-native-kubernetes-usage-report-2022), many respondents appear to utilize Kubernetes' HA architecture for highly secure, data-sensitive applications.

In this section, we will go over some of the best practices for deploying HA apps in Kubernetes. These guidelines build upon what we have seen in *Chapter 5, Creating and Implementing Updates on Multi-Node Raspberry Pi Kubernetes Clusters*.

As you may be aware, deploying a basic app setup in Kubernetes is a piece of cake. Trying to make your application available and fault-tolerant, on the other hand, implies a slew of challenges and problems. In general, implementing HA in any capacity requires the following:

- **Determining your application's intended level of availability**: The allowable level of downtime varies by application and business objectives.

- **A redundant and reliable control plane for your application**: The control plane manages the cluster state and contributes to the availability of your applications to users.

- **A redundant and reliable data plane for your application**: This entails duplicating the data across all cluster nodes.

There are a lot of considerations and decisions to make while deploying Kubernetes' HA that might have an impact on the apps and how they operate and consume storage resources. Here, we will look at some of the considerations to make:

- **Use of replicas**: Use replicas instead of pods to deploy HA apps. Using replicas ensures that your application is operating on a consistent set of pods at all times. For the application to be declared minimally accessible, it must have at least two replicas.

- **Review your update strategy**: The default deployment update strategy involves reducing the number of old and new ReplicaSet pods with a `Ready` state to 75% of their pre-update level. As a result, during the update, an application's compute capacity may drop to 75% of its normal level, resulting in a partial failure (degraded application performance). The `RollingUpdate.maxUnavailable` parameter lets you choose the maximum percentage of pods that can go down during an upgrade. As a result, either ensure that your application operates properly, even if 25% of your pods are unavailable, or reduce the `maxUnavailable` option. Based on the application needs, other deployment strategies, such as blue-green and canary, among others, can also be evaluated for a much better alternative to the default strategy.

- **Right sizing of the nodes**: The maximum amount of RAM that can be allocated to pods is determined by the size of the nodes. For production clusters, the size of the nodes is large enough (2.5 GB or more) that they can absorb the workload of any crashed nodes.

- **Node pools for HA**: Node pools with production workloads should contain at least three nodes to provide HA. This allows the cluster to distribute and schedule work on other nodes if one becomes unavailable.

- **Set requests and limits**: To keep your cluster running efficiently, define the `request` and `limit` objects in your application spec for all deployments:

    A. **Requests**: Specifies how much of a resource (such as CPU and memory resources) a pod may require before it is scheduled on a node. The pod will not be scheduled if the node lacks the required resources. This keeps pods from being scheduled on nodes that are already overburdened.

B. **Limits**: Specifies how many resources (such as CPU and RAM) a pod is permitted to use on a node. This stops pods from potentially slowing down the operation of other pods.

- **Set pod disruption budgets**: To avoid interruptions to production, such as during cluster upgrades, you can configure a pod disruption budget, which restricts the number of replicated pods that can be down at the same time.

- **Ensure a cluster is upgraded**: Ensure to take advantage of the latest features, security patches, and stability improvements.

- **Avoid single points of failure**: Kubernetes enhances dependability by providing repeating components and ensuring that application containers can be scheduled across various nodes. For HA, use anti-affinity or node selection to help disperse your applications across the Kubernetes cluster. Based on labels, node selection allows you to specify which nodes in your cluster are eligible to run your application. Labels often describe node attributes such as bandwidth or specialized resources such as GPUs. For example, to properly commit mutations to data, Apache ZooKeeper requires a quorum of servers. Two servers in a three-server ensemble must be healthy for writes to succeed. As a result, a resilient deployment must make sure that servers are distributed across failure domains.

- **Use liveness and readiness probes**: By default, Kubernetes will transfer traffic to application containers instantaneously. You may improve the robustness of your application by configuring health checks to notify Kubernetes when your application pods are ready to receive traffic or have become unresponsive.

- **Use initContainers**: Before running the primary containers, `startupProbe` or `readinessProbe`, you can use `initContainers` to check for external dependencies. Changes to the application code are not required for `initContainers`. It is not necessary to embed additional tools in order to utilize them to examine external dependencies in application containers.

- **Use plenty of descriptive labels**: Labels are extremely powerful since they are arbitrary key-value pairs and enable you to logically organize all your Kubernetes workloads in your clusters.

- **Use sidecars for proxies and watchers**: Sometimes, a set of processes is required to communicate with another process. However, you do not want all of these to operate in a single container but rather in a pod. This is also the case when you are running a proxy or a watcher on which your processes rely. For example, with a database on which your processes rely, the credentials would not be hardcoded onto each container. Instead, you can deploy the credentials as a proxy inside a sidecar that handles the connection securely.

- **Automate your CI/CD pipeline and avoid manual Kubernetes deployments**: Because there may be numerous deployments per day, this strategy saves the team considerable time by eliminating manual error-prone tasks.

- **Use namespaces to split up your cluster**: For example, you can construct Prod, Dev, and Test namespaces in the same cluster, and you can also use namespaces to limit the number of resources so that one defective process does not consume all of the cluster resources.

- **Monitoring the control plane**: This helps in the identification of issues or threats within the cluster and increases latency. It is also advised to employ automated monitoring tools rather than managing alerts manually.

To recap, we've gone through some of the best practices to optimize your Kubernetes environment.

Summary

In this chapter, we looked at how to set up an HA MicroK8s Kubernetes cluster using the stacked cluster HA topology. We utilized the three nodes to install and configure MicroK8s on each of them, as well as simulating node failure to see whether the cluster could tolerate component failures and still continue to function normally.

We discussed some of the best practices for implementing Kubernetes applications on your production-ready cluster. We also covered the fact that MicroK8s' HA option has been simplified and enabled by default.

HA is a vital feature for organizations looking to deploy containers and pods that can deliver the kind of reliability required at scale. We also recognized the value of Canonical's lightweight Dqlite SQL database, which is used to provide HA clustering. By embedding the database into Kubernetes, Dqlite reduces the cluster's memory footprint and eliminates process overhead. For IoT or Edge applications, this is critical.

In the next chapter, we'll look at how to use Kata Containers, a secure container runtime, to provide stronger workload isolation by leveraging hardware virtualization technology.

14

Hardware Virtualization for Securing Containers

In the previous chapter, we looked at how to create a highly available MicroK8s Kubernetes cluster using the stacked cluster **high-availability** (**HA**) topology. We have used the three nodes to install and configure MicroK8s on each of them, as well as simulating node failure to see whether the cluster could withstand component failures and still work normally. We've also gone over some best practices for deploying Kubernetes applications on a production-ready cluster. We noticed that MicroK8s' HA option has also been streamlined and is now activated by default.

Container technologies have dominated the industry in recent years and have become the de facto standard for building modern IT infrastructure. They are frequently preferred over standard **virtual machines** (**VMs**) due to their lightweight design and bare-metal-like performance. However, security and isolation are two of the most common adoption issues (refer to the recent Kubernetes and cloud-native operations report, published in 2022, at `https://juju.is/cloud-native-kubernetes-usage-report-2022`). In this chapter, we'll look at how to use Kata Containers to create a secure container runtime and leverage hardware virtualization technology to give better workload isolation.

Before understanding what Kata Containers is, let us review how containers operate and how they relate to virtualization technology. A container is more like a **VM** that allows the packaging of software and all of its dependencies into a single entity that can execute in any supported environment.

VMs, on the other hand, are larger and require longer to set up. Containers have a substantially lower footprint than VMs and are thus much faster to set up (and tear down). Containers, unlike VMs, which keep entire copies of the operating system, only share the host system's operating system kernel.

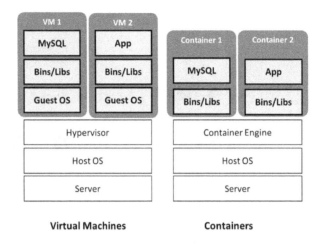

Figure 14.1 – VMs versus containers

The container runtime is the bridging software that allows a host system to separate its resources for containers, tear down container images, and manage container life cycles. Every node on the Kubernetes cluster must have a container runtime installed.

Canonical MicroK8s has made it easy to enable Kata Containers (a container runtime), which can greatly improve the security and isolation of your container operations, with just a single command. It combines the advantages of a hypervisor, such as increased security, with Kubernetes' container orchestration capabilities. In this chapter, we're going to cover the following main topics:

- Overview of Kata Containers
- Enabling the Kata add-on and running a sample application
- Container security best practices

Overview of Kata Containers

The **Open Container Interface** (**OCI**) is a Linux Foundation initiative that aims to establish principles, standards, and specifications for Linux containers. The OCI runtime specifications are primarily concerned with container life cycle management and configuration for multiple systems, including Linux, Windows, and Solaris. Low-level runtimes are container runtimes that comply with the OCI specification. Container creation and management are primarily the responsibility of low-level container runtimes. Designed by Docker, runC is an example of low-level container runtime and the standard for low-level container runtimes.

Low-level runtimes are native runtimes, which means they run containerized processes on the host kernel. There are also a few sandboxed and virtualized runtimes that provide improved process isolation by not running them on the host kernel. Kata Containers is one of the virtualized runtimes. To run containerized processes, these runtimes use a VM interface that behaves similarly to containers, but with the workload isolation and security benefits of VMs.

The Docker runtime was the default container runtime when Kubernetes was initially published. As the platform developed, so did the need to support different runtimes.

The **Container Runtime Interface** (**CRI**) was launched to make Kubernetes more runtime agnostic. It's a high-level specification that's mostly focused on container orchestration. Unlike the OCI, the CRI handles extra aspects of container administration such as image management, snapshots, and networking while leaving container execution to an OCI-compliant runtime (such as runC).

Kata Containers (`https://katacontainers.io/`) is an open source project that seeks to create a secure and OCI-compliant container runtime that improves the security and isolation of container workloads by encapsulating each one in a lightweight VM and employing hardware virtualization. Every VM has its own kernel that it uses.

As depicted in *Figure 14.2*, traditional containers employ runC as a container runtime, which relies on kernel features such as cgroups and namespaces to achieve isolation with the shared kernel; however, Kata Containers leverages hardware virtualization to isolate containers in their own lightweight VM as follows:

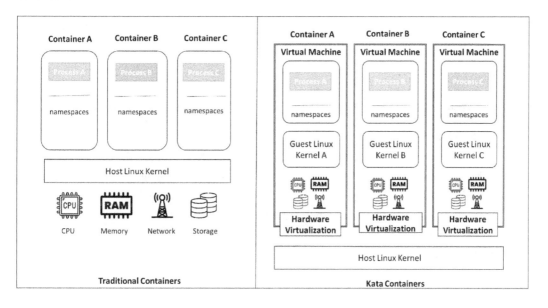

Figure 14.2 – Traditional containers versus Kata Containers

Kata Containers has a number of advantages over standard VMs, including the ability to effortlessly integrate with existing container orchestration technologies such as Kubernetes. Native Kubernetes capabilities such as auto-scaling and rolling updates are still available while you're launching VMs. This enables the benefits of virtualization technology to be combined with container orchestration capabilities. In the following section, we can look at how the Kata containers are instantiated using kata-runtime.

How Kata Containers works

When a Kubernetes cluster is configured with a high-level runtime, such as containerd or CRI-O, a container runtime shim is installed to allow smooth communication between the CRI (containerd or CRI-O) and a low-level container runtime, such as runC (the default runtime), and this low-level container runtime is responsible for running the containers in the pod.

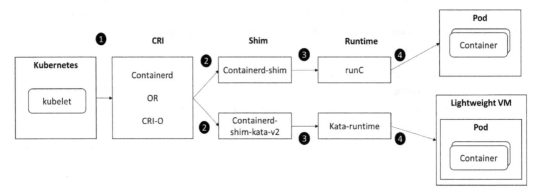

Figure 14.3 – Working of Kata Containers

The steps for creating Kata containers with segregated kernel and namespace are as follows:

1. Kubernetes is configured with a high-level container runtime such as **containerd** or **CRI-O**.

2. A container runtime shim (**containerd-shim**) acts as a bridge between the CRI (containerd or CRI-O) and a low-level container runtime, such as runC (the default runtime), for smooth communication.

3. The low-level container runtime (such as runC or kata-runtime) takes care of running the containers in the pod.

4. Kata Containers uses a runtime class (kata-runtime) to run containers in a separate kernel and namespace.

Containers can be run in a lightweight VM using **containerd** or **CRI-O** once Kata Containers is enabled on the Kubernetes cluster with kata-runtime. `containerd-shim-kata-v2`, a new shim that acts as a bridge between containerd and `kata-runtime`, Kata Containers' runtime class should also be enabled for running containers in a separate kernel and namespace.

In a nutshell, Kata is a container runtime that provides greater isolation between containers while maintaining the performance and efficiency of other runtimes. The following are some of its prominent features:

- **Security**: It runs in a dedicated and isolated kernel and can be easily integrated with containerd or any other container runtime. It also supports several hypervisors such as QEMU, Cloud Hypervisor, and Firecracker.

- **Compatibility with Docker and Kubernetes**: By providing kata-runtime as a container runtime, it works easily with Docker and Kubernetes.

- **Performance**: It has the same consistency as any other Linux container but with more isolation. It also supports AMD64, ARM, IBM pSeries, and IBM zSeries platforms.

- **Simplified**: No need for nested containers inside VMs or sacrificing container speed.

Apart from virtualized runtimes or Kata Containers, there are various techniques to isolate containers (refer to `https://thenewstack.io/how-to-implement-secure-containers-using-googles-gvisor/`), each with its own set of attributes that will suit certain applications. The appropriate one for your apps is a crucial aspect of your container security architecture.

Now that we've grasped the fundamentals of how Kata Containers works, we can move on to the following step of enabling the Kata add-on and running a sample application.

Enabling the Kata add-on and running a sample application

In this section, we can go through the process of enabling the Kata add-on in your MicroK8s Kubernetes cluster. Then, to demonstrate Kata's capabilities, we'll deploy a sample application.

> **Note**
> We'll be using an Ubuntu VM for this section. The instructions for setting up the MicroK8s cluster are the same as in *Chapter 5, Creating and Implementing Updates on Multi-Node Raspberry Pi Kubernetes Clusters*.

Step 1 – Enabling the Kata add-on

Starting with MicroK8s v1.24, you must issue the `enable community` command to enable the community add-ons repository.

Use the following command to enable the `community` repository:

```
microk8s enable community
```

It will take some time to finish activating the add-on; the following command execution output shows that the `community` repository has been successfully enabled:

```
$ microk8s enable community
Infer repository core for addon community
Cloning into '/var/snap/microk8s/common/addons/community'...
done.
Community repository is now enabled
$ 
```

Figure 14.4 – Enabling the community repository

Now that we have the `community` repository enabled, we can move on to the following step of enabling the Kata add-on.

Use the following command to enable the Kata add-on:

```
microk8s enable kata
```

The following command execution output indicates that the Kata add-on is being enabled:

```
$ microk8s enable kata
Infer repository community for addon kata
Installing kata-containers snap
kata-containers 2.4.1 from Kata Containers (katacontainers√) installed
Restarting containerd
Warning: node.k8s.io/v1beta1 RuntimeClass is deprecated in v1.22+, unavailable in v1.25+
runtimeclass.node.k8s.io/kata created


To use the kata runtime set the 'kata' runtimeClassName, eg:

kind: Pod
metadata:
  name: nginx-kata
spec:
  runtimeClassName: kata
  containers:
  - name: nginx
    image: nginx

$ 
```

Figure 14.5 – Enabling the Kata add-on

From the preceding command execution output, we can see that `kata runtimeClassName` (kata) is added, which allows us to specify which workloads should be launched in Kata Containers.

> **Note**
>
> The `--runtime-path` parameter can also be used to specify the location where the Kata runtime is installed.
>
> Use the following command to enable the Kata add-on with the runtime path:
>
> ```
> microk8s enable kata --runtime-path=<<kata-runtime-binary-path>>
> ```

Before we move on to the following step, let's make sure that the Kata add-on has been activated successfully using the `microk8s status` command as follows:

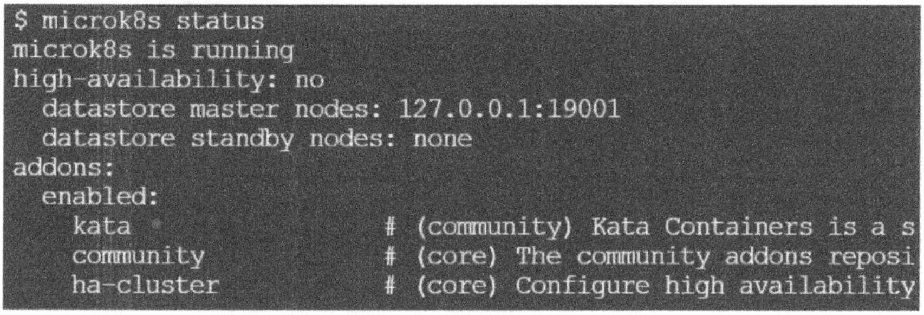

```
$ microk8s status
microk8s is running
high-availability: no
    datastore master nodes: 127.0.0.1:19001
    datastore standby nodes: none
addons:
    enabled:
      kata              # (community) Kata Containers is a s
      community         # (core) The community addons reposi
      ha-cluster        # (core) Configure high availability
```

Figure 14.6 – The Kata add-on is enabled

Now that the Kata add-on has been enabled, we may deploy a sample nginx application in the following step.

> **Note for using Kata Containers on multi-node clusters**
>
> `microk8s enable kata` must be executed on each node in a multi-node cluster in order for the Kata runtime to be enabled on the desired nodes.

Step 2 – Deploying a sample application

In this step, we will be deploying the following sample nginx application deployment manifest, which uses the Kata runtime:

```
apiVersion: v1
kind: Pod
metadata:
  labels:
    app: kata
  name: nginx-kata
spec:
  runtimeClassName: kata
```

```
containers:
  - name: nginx
    image: nginx
```

Use the following command to create a sample nginx deployment:

```
kubectl apply -f kata-nginx.yaml
```

The following command execution output indicates that there is no error in the deployment and in the following steps, we can ensure that the pods are created as follows:

```
$ kubectl apply -f kata-nginx.yaml
pod/kata-nginx created
$
```

Figure 14.7 – Sample nginx application deployed

Now that the deployment is successful, let's use the following `kubectl` command to check whether the pods are in a `Running` state:

```
$ kubectl get pods
NAME          READY    STATUS    RESTARTS    AGE
nginx-kata    1/1      Running   0           12s
$
```

Figure 14.8 – Checking whether the pods are in a Running state

The fact that `nginx-kata` is now in the `Running` state means that containers are running in a lightweight VM utilizing the containerd runtime. It used `containerd-shim-kata-v2`, which acts as a bridge between containerd and kata-runtime—a runtime class that comes with Kata Containers that allows containers to run in their own kernel and namespace.

Now that we've seen how easy is to enable the Kata add-on and run a sample application, let's move on to the following section where we discuss the best practices for running containers.

Container security best practices

Containers provide a lot of advantages, but they also have some security issues that might be tough to solve. Because of the enormous number of containers based on many different underlying images, each of which could potentially have vulnerabilities, containers create a wider attack surface than traditional workloads.

Another important consideration is the kernel architecture shared by containers. Securing the host is insufficient to ensure security. You must also keep secure configurations to limit container permissions and ensure effective container isolation. For example, a container with an exploitable vulnerability, exposed metadata, and incorrect credentials configuration could jeopardize your entire infrastructure.

We'll go through some of the most important factors to consider when running containers.

Utilizing DevSecOps

The seamless integration of security testing and protection across the software development and deployment life cycle is referred to as DevSecOps. You may scan your code for defects or possibly vulnerable code before shipping or even building your application. There are various **Static Application Security Testing (SAST)** tools for application code, such as **SonarQube**, which provides vulnerability scanners for various languages and detects vulnerabilities based on rules, linters, and so on. You can use them on the development workstation, but including code scanning tools in the CI/CD workflow ensures a minimal degree of code quality. For example, if some checks fail, you can deny pull requests by default.

Also, remove any components that your application doesn't require. For example, remove the sed and awk binaries, which are installed by default on any UNIX system. This can assist you in lowering the attack surface.

Scanning external vulnerabilities via dependency scanning

External dependencies, such as third-party libraries or frameworks that are used in your application, may contain flaws and vulnerabilities. Any application build process should incorporate dependency scanning as a best practice.

A vulnerability database (such as NVD) can also be matched with your application dependencies using package management tools such as npm, maven, go, and others to produce helpful alerts/warnings.

Analyzing container images using image scanning tools

Analyze your container images with an image scanner. The image scanning tool will find vulnerabilities in the operating system packages provided by the container image base distribution (rpm, dpkg, apk, and so on). It will discover vulnerabilities in package dependencies for Java, Node.js, Python, and other languages.

Automating and enforcing image scanning is simple. It can be integrated into your CI/CD pipelines, triggered when new images are sent to a registry to ensure that non-compliant images are no longer permitted to run.

Enforcing image content trust

If you aren't making the image from scratch, you should pick images that are reliable. Anyone can utilize public image repositories such as Docker Hub, and they may contain viruses or misconfigurations.

Container image integrity can also be enforced by using Docker Notary or a comparable service to add digital signatures to the image, which can then be validated in the container runtime.

Securing registries

Container images are often saved in either private or public registries. It's crucial to keep these registries safe so that all team members and collaborators can use images that are as secure as possible.

If you have your own private registry, you must set up access controls that specify who can and cannot access and publish images. Access control is a fundamental security measure that can prevent unauthorized parties from altering, publishing, or deleting your images.

Securing your host

It's just as critical to secure your host as it is to secure your containers. The host where the containers operate is often made up of an operating system with a Linux kernel, a collection of libraries, a container runtime, and various background services and helpers. Any of these components could be insecure or misconfigured, allowing unauthorized access to running containers or a **Denial of Service (DoS)** attack.

For example, difficulties with the container runtime itself, such as a DoS attack that prohibits the creation of new containers in a host, can have an impact on your operating containers. You could use host scanning utilities to identify known security holes in the host's container runtime, services, standard libraries such glibc, and the kernel (quite similar to what image scanning does for a container image).

Securing your runtime

The following are some best practices for ensuring runtime security:

- *Create separate virtual networks for your containers*: This adds a layer of isolation that can help limit the attack surface.
- *Use the principle of least privilege*: Only allow connectivity between containers that genuinely require it.
- *Only expose the ports required by the application*: Except for SSH, do not expose any additional ports. Apply this idea to both containers and underlying computers.
- *Use TLS to secure service communication*: This method encrypts traffic and ensures that only authorized endpoints are allowed.

- *Use the Docker image policy plugin*: This plugin prevents any process from retrieving images that have not previously been allow-listed.

Reviewing container privileges

The scope of a vulnerability exploited inside a container is largely determined by the container's privileges and isolation from the host and other resources. The existing and prospective vulnerabilities can be mitigated by using runtime settings in the following ways:

- Run the container as a user, not as a root. Use randomized UIDs if possible.

- Docker and Kubernetes both allow for the removal of capabilities and the disabling of privileged containers. Seccomp and AppArmor can limit the types of actions that a container can execute.

- To avoid a container taking all of the memory or CPUs and starving other apps, use resource limits.

- Examine shared storage or volumes on a regular basis, paying special attention to the host path and sharing the filesystem from the host.

- Pod Security Policies can be used to create guardrails in your cluster and prevent misconfigured containers.

Using real-time event and log auditing

Threats to container security can be detected by evaluating aberrant activity and auditing several sources of logs and events. The following are some examples of sources of events:

- Logs from the host and Kubernetes
- Container calls to the operating system

Use tools (such as Falco and Sysdig Secure) that can track the system calls that are made and send out alerts if anything unusual happens. It should come with a pre-configured library of rules as well as the option to write your own using a simple syntax. It should also be able to monitor the Kubernetes audit log.

Monitoring resource usage

Excessive resource utilization (CPU, RAM, and network), a rapid drop in available disc space, an above-average error rate, or increased latency could all be signs that something is wrong with your system.

Collect metrics in the same way as Prometheus (refer to *Chapter 8, Monitoring the Health of Infrastructure and Applications*) does. Set up alerts to be notified as soon as the data exceeds the predicted thresholds. Use useful dashboards to track the evolution of metrics and see how they relate to other metrics and events in your system.

Common security misconfigurations and remediation

Incorrectly configured hosts, container runtimes, clusters, resources, and so on may give an attacker a way to increase their privileges and move upward.

Learn how to spot configuration errors, why they're problematic, and how to correct them by using benchmarks, best practices, and hardening guidelines. The **Center for Internet Security (CIS)** (`https://www.cisecurity.org/benchmark/kubernetes`) is the most authoritative source of information that provides free benchmarks for a variety of situations, and anyone and any firm can contribute their expertise.

The easiest method to ensure security is to automate as much as possible. There are a number of tools, such as kube-bench (`https://github.com/aquasecurity/kube-bench`), most of which are based on static configuration analysis, allow you to evaluate configuration parameters at various levels, and provide recommendations on how to change them.

To recap, security controls that safeguard containers and the underlying infrastructure should be implemented and maintained as part of container security. Integrating security into the development pipeline can ensure that all components are protected from the beginning of their development phase to the end of their life cycle.

Summary

We looked at how to use Kata Containers to build a secure container runtime and how to employ hardware virtualization technology to improve workload isolation. We have also looked at how to enable the Kata add-on and run a sample application.

We discussed best practices for establishing container security on your production-grade cluster. We also noticed how the MicroK8s add-on option has made it simpler to activate Kata Containers, which can dramatically improve the security and isolation of your container operations.

With Kata Containers maturing into a production-ready container runtime and subsequent uptake, there is a great opportunity to improve the hosted build and development environment approach to address the noisy neighbor problem and handle unique and privileged requirements without affecting current host settings or policies.

In the following chapter, we will continue our next use case of implementing strict confinement for isolated containers.

15

Implementing Strict Confinement for Isolated Containers

In the previous chapter, we looked at how to build secure containers using Kata Containers and how to improve workload isolation with hardware virtualization technology. We also discussed the best practices for securing your production-grade cluster with containers. The MicroK8s add-on option has also made it easier to activate Kata Containers, which can significantly increase the security and isolation of your container operations.

In this chapter, we will take a look at another approach to isolation using snap confinement options to run containers in complete isolation, meaning no access to files, networks, processes, or any other system resource without requesting specific access via an interface. Confinement models describe how much access a particular snap has to a user's machine. There are currently three choices available, as follows:

- The *strict* confinement level operates in complete isolation, with access limited to a level that is always regarded as safe. As a result, without requesting particular access via an interface, strictly limited snaps cannot access your files, network, processes, or any other system resource.

- The *classic* confinement level is similar to conventional Linux packages and can access a system's resources.

- The *devmode* confinement level runs in a limited environment with full access to system resources and generates debug output to locate unidentified interfaces. This is specifically designed for snap creators and developers.

As we discussed in *Chapter 2, Introducing MicroK8s*, MicroK8s is a snap, and we employed the classic confinement model throughout this book. The interface of each snap is carefully chosen by the author to enable specialized access to a resource in accordance with the snap's requirements. For example, network access, desktop access, and audio access are all provided by common interfaces.

In this chapter, we're going to cover the following main topics:

- Overview of Snap, Snapcraft, and Ubuntu Core
- Setting up Ubuntu Core on a Raspberry Pi board
- Setting up MicroK8s on Ubuntu Core
- Deploying a sample containerized application

Overview of Snap, Snapcraft, and Ubuntu Core

Before going into detail about how strict confinement snaps offer isolation to applications, we will delve into a little bit of history on how embedded Linux development was handled before the advent of snaps.

Traditionally, getting software to embedded Linux systems has proven difficult. There were different Linux packaging formats (RPM, DEB, and so on) and there is no standardization of formats. Furthermore, software packages frequently necessitate sophisticated code to manage installation and updates that are incompatible with one another, have unmet dependencies, or write to the entire system.

Snaps were envisioned to answer the concerns of embedded Linux developers seeking a secure environment and precise configuration to run their applications. They allow software publishers and developers to manage the binary that is supplied and the exact version that their users have access to.

Snaps are easy to create, build, and deploy as compared to standard Linux software distribution. Snaps get automatically updated **over the air** (**OTA**) and via deltas, keeping the functionality of an embedded Linux device always fresh and reducing the risk of breaking. Snaps are compatible with all major Linux distributions and may be used on any type of device from desktop to cloud and IoT devices.

Snapcraft (`https://snapcraft.io/`) is a framework for creating and distributing snaps by bringing together different components of the application into a single, cohesive bundle. Developers submit their snaps to a central repository known as Snap Store—a universal app store that allows users to publish, browse, install, distribute, and deploy apps in the cloud, on desktops, and on IoT devices from any Linux distribution.

The Ubuntu Core embedded OS (`https://ubuntu.com/core`) is built on snaps and is free and open source. In Ubuntu Core, everything is a snap. Even the kernel is a snap. In Ubuntu Core, only snaps that use the strict confinement model can be installed. It is a cutting-edge new operating system (OS) designed from the ground up with zero-trust security in mind. It efficiently decouples the base system and OS from the installed apps by containerizing the Linux kernel and runtime environments. Containerization allows you to separate and provide lockdown functionality, with applications running in a security sandbox by default (kernel features such as AppArmor, seccomp, security policies, and device permissions are leveraged).

MicroK8s and Ubuntu Core share characteristics including self-healing, high availability, automatic OTA updates, reliability, and security. Running MicroK8s on Ubuntu Core provides Kubernetes with the benefits of a solid computing foundation. Furthermore, combining Ubuntu Core and MicroK8s offers a streamlined, embedded Kubernetes experience for IoT and Edge applications, with a small footprint and performance efficiency optimizations.

In the next section, we will go over the procedures for setting up a Kubernetes Raspberry Pi cluster with snap strict confinement.

Setting up Ubuntu Core on a Raspberry Pi board

Now that we are clear on the snap confinement concepts, we will delve into the steps of creating a Kubernetes Raspberry Pi cluster that uses snap strict confinement.

What we are trying to achieve

We'll list down the steps that we're seeking to work through in this section as follows:

1. Setting up an Ubuntu Core image to SD card
2. Creating an Ubuntu SSO account
3. Generating an SSH key pair
4. Booting Ubuntu Core on Raspberry Pi

The Raspberry Pi cluster that we will build in this step is depicted in *Figure 15.1* as follows:

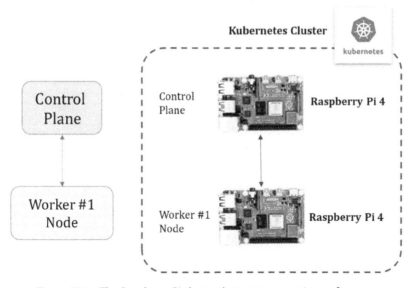

Figure 15.1 – The Raspberry Pi cluster that uses snap strict confinement

Now that we know what we want to do, let's look at the requirements.

Requirements

The following are the prerequisites for building the Ubuntu Core Raspberry Pi Kubernetes cluster:

- A microSD card (4 GB minimum; 8 GB recommended)
- A computer with a microSD card drive
- Raspberry Pi 2, 3, or 4 (1 or more)
- A micro-USB power cable (USB-C for the Pi 4)
- A Wi-Fi network or an Ethernet cable with an internet connection
- (Optional) A monitor with an HDMI interface
- (Optional) An HDMI cable for Pi 2 and 3 and a micro-HDMI cable for Pi 4
- (Optional) A USB keyboard

Now that we've established the requirements, we'll move on to the step-by-step instructions on how to create a Kubernetes Raspberry Pi cluster that uses snap strict confinement.

Step 1 – Setting up an Ubuntu Core image to an SD card

The first step is to install an Ubuntu Core image to the microSD card. To do that, we will be using the **Raspberry Pi Imager tool** to install an OS image to a microSD card that can then be used with Raspberry Pi.

Download and install **Raspberry Pi Imager** from the Raspberry Pi website on a computer equipped with an SD card reader.

As depicted in *Figure 15.2*, run Raspberry Pi Imager with the microSD card and open the **CHOOSE OS** menu as follows:

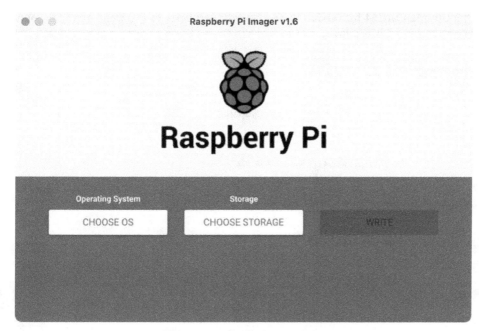

Figure 15.2 – Raspberry Pi Imager

From the OS menu, choose **Other general purpose OS** from the options listed, as follows:

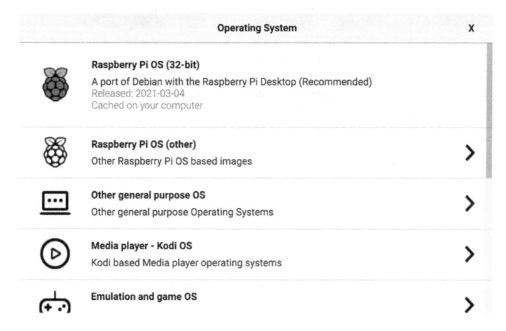

Figure 15.3 – Raspberry Pi Imager OS options

Choose the **Ubuntu Core 64-bit version** that works with Raspberry Pi 2,3, 3, and 4 from the options listed (refer to *Figure 15.4*) as follows:

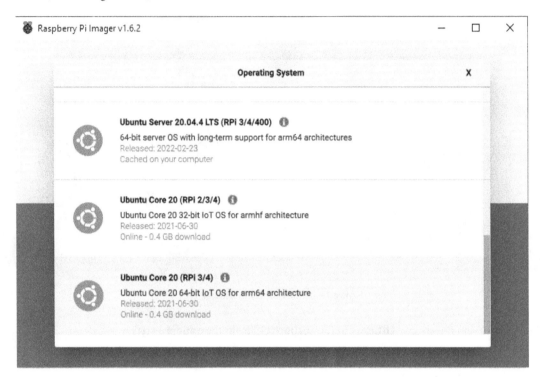

Figure 15.4 – Choose Ubuntu Core 64-bit version

As depicted in *Figure 15.5*, open the **Storage** menu after selecting **Ubuntu Core 64-bit image**. Choose the micro SD card that you've inserted as follows:

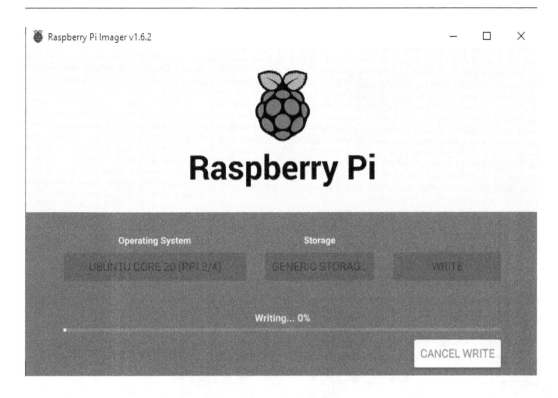

Figure 15.5 – Raspberry Pi Imager write operation

Finally, click on **Write** to start the operation and Raspberry Pi Imager will wipe your micro SD card data; you will be prompted to confirm this procedure.

Post confirmation, Raspberry Pi Imager will start flashing OS images to the micro SD card. It will take a while to finish.

Once finished, continue with the creation of an Ubuntu Single Sign-On (SSO) account.

Step 2 – Creating an Ubuntu SSO account

An Ubuntu **SSO** account needs to be created so that Secure shell (SSH) public keys can be stored and linked to an email address. This allows Ubuntu Core devices to only permit SSH connections from the devices that have public keys that match those in your SSO account.

Go to `https://login.ubuntu.com/` and fill in the relevant information, and after the SSO account has been created, generate the SSH key pair as explained in the following section.

Step 3 – Generating an SSH key pair

SSH, as we know, is a popular way to connect to remote Linux servers. The authentication process involves the pairing of a private local key with a public remote key, which is used to secure communication from your device to Linux servers that are hosting the application.

Using the free and open source OpenSSH software included in Windows 10, SSH keys can be generated. SSH keys can also be generated through the PuTTYgen utility, which has support for various platforms. We're going to use the built-in Windows OpenSSH client for the following steps.

From the PowerShell window, enter the following command:

```
ssh-keygen
```

The first step asks where you would like to save the key, and you can accept the default answer by pressing Return, as shown in the command execution output in *Figure 15.6*. The passphrase is requested in the second step. When a passphrase is entered, it is required to use the passphrase each time the key is accessed. The passphrase is optional; pressing Return twice will create a key pair without requiring a passphrase, as follows:

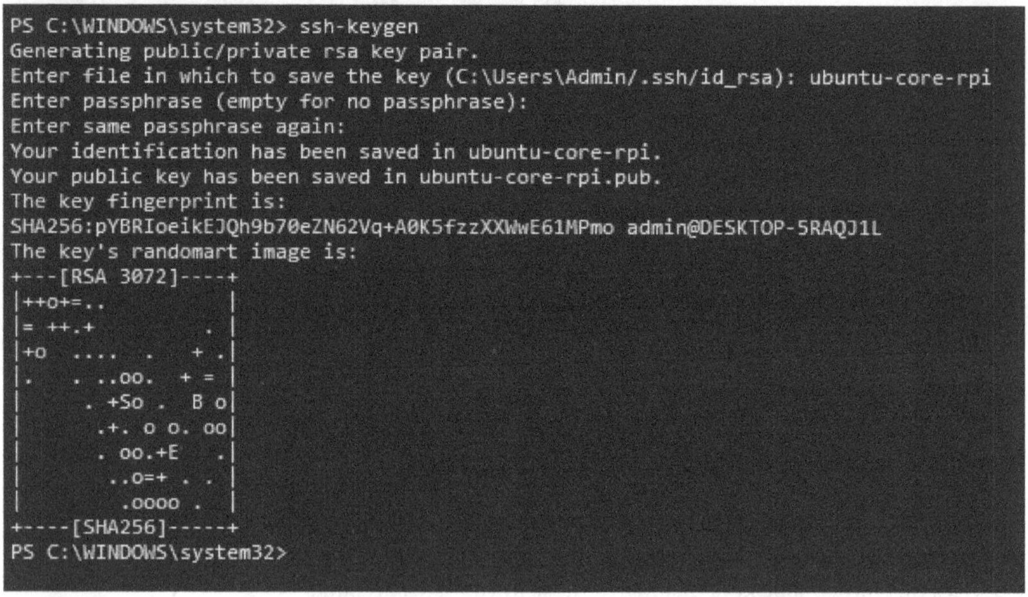

```
PS C:\WINDOWS\system32> ssh-keygen
Generating public/private rsa key pair.
Enter file in which to save the key (C:\Users\Admin/.ssh/id_rsa): ubuntu-core-rpi
Enter passphrase (empty for no passphrase):
Enter same passphrase again:
Your identification has been saved in ubuntu-core-rpi.
Your public key has been saved in ubuntu-core-rpi.pub.
The key fingerprint is:
SHA256:pYBRIoeikEJQh9b70eZN62Vq+A0K5fzzXXWwE61MPmo admin@DESKTOP-5RAQJ1L
The key's randomart image is:
+---[RSA 3072]----+
|++o+=..          |
|= ++.+         . |
|+o   ....   .  + .|
|.    . ..oo.  + = |
|      . +So .  B o|
|      .+. o o. oo|
|      . oo.+E   .|
|       ..o=+ . . |
|         .oooo . |
+----[SHA256]-----+
PS C:\WINDOWS\system32>
```

Figure 15.6 – SSH key generation

The private key and the public key can be found in the same folder once the process is complete, as follows:

Name	Date modified
ubuntu-core-rpi	6/20/2022 10:46 AM
ubuntu-core-rpi.pub	6/20/2022 10:46 AM

Figure 15.7 – Private and public keys generated

We can now use the generated public and private keys for Ubuntu Core installation. The following step is to add the public key to an Ubuntu SSO account so that it could be used to permit connections from the devices that have public keys.

From the Ubuntu SSO account login, under the **SSH keys** section, copy the contents of the ubuntu-core-rpi.pub (public key) file to import the public key as follows:

Figure 15.8 – Import SSH keys

The following screenshot shows that the public key has been imported successfully:

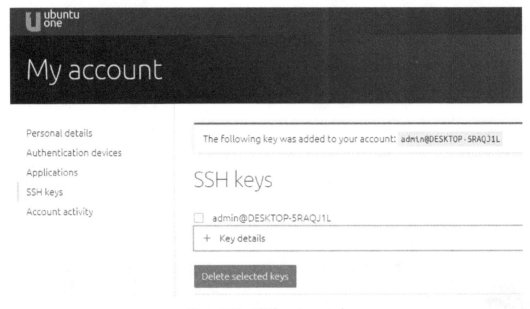

Figure 15.9 – SSH keys imported

Now that we have imported SSH keys to the Ubuntu SSO account, the following step is to power Raspberry Pi and boot Ubuntu Core.

Step 4 – Booting Ubuntu Core on Raspberry Pi

Extract the SD card from your laptop and insert it into Raspberry Pi. Before powering the Pi, connect an HDMI screen and a USB keyboard. Power on the Pi and you will be able to see the boot process on the screen. It typically takes less than 5 minutes to complete the booting process.

When the boot process is complete, you will see instructions for configuring the network and creating an administrator account on Ubuntu Core. In this configuration, you will be able to configure Wi-Fi settings, and the following step will require you to provide the email address associated with your SSO account. Once configured, the device will automatically update and, if necessary, restart.

After you provide your email and the Pi connects to your account, you will be able to use any SSH client, for example, PuTTY, to connect to your Pi.

Success! You are now connected to Ubuntu Core running on your Raspberry Pi.

We've finished configuring the settings and we are ready to go on to the following step of installing and configuring the MicroK8s snap with strict confinement.

Setting up MicroK8s on Ubuntu Core

SSH into your control plane node and install the latest version of the MicroK8s snap with strict confinement as in the following command:

```
sudo snap install microk8s --channel=latest/edge/strict
```

The following is the output after the execution of the preceding command, which confirms that the MicroK8s snap with strict confinement was successfully configured:

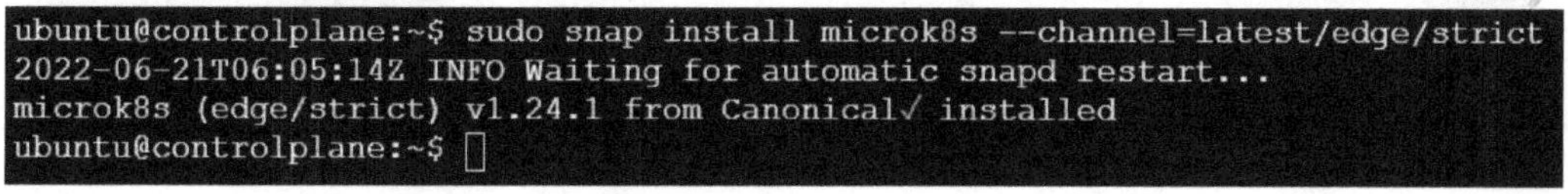

```
ubuntu@controlplane:~$ sudo snap install microk8s --channel=latest/edge/strict
2022-06-21T06:05:14Z INFO Waiting for automatic snapd restart...
microk8s (edge/strict) v1.24.1 from Canonical√ installed
ubuntu@controlplane:~$ []
```

Figure 15.10 – Successful MicroK8s snap installation

Now that we have installed the MicroK8s snap, let's run the `microk8s status` command to verify its running state as follows:

```
microk8s status
```

The following is the output after the execution of the preceding command, which confirms that the MicroK8s snap with strict confinement is running successfully:

```
ubuntu@controlplane:~$ microk8s status
microk8s is running
high-availability: no
  datastore master nodes: 127.0.0.1:19001
  datastore standby nodes: none
addons:
  enabled:
    ha-cluster            # Configure high av
  disabled:
    dashboard             # The Kubernetes da
    dashboard-ingress     # Ingress definitio
```

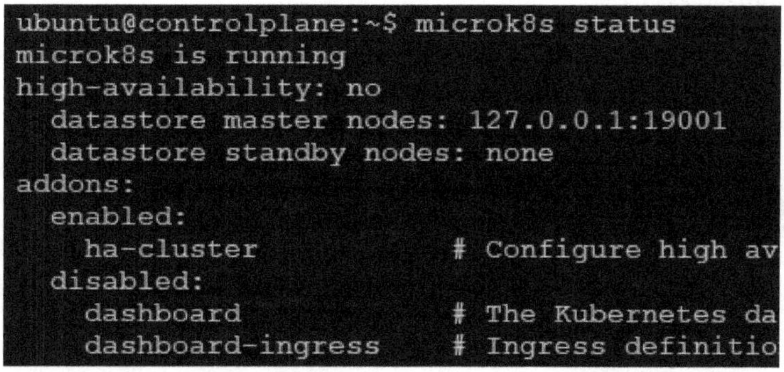

Figure 15.11 – MicroK8s snap is running

Strict confinement *locks down* the apps in the snap using Linux kernel security capabilities. Access will be extremely constrained for a highly contained application without any stated interfaces. MicroK8s running successfully indicates that all necessary interfaces are specified and application access requirements are met.

To view the MicroK8s snap interfaces, use the following command:

```
snap connections microk8s
```

The following is the output after the execution of the preceding command, which lists interfaces of the MicroK8s snap:

```
root@master:~# snap connections microk8s
Interface          Plug                             Slot                  Notes
account-control    microk8s:account-control         :account-control      -
cifs-mount         microk8s:cifs-mount              -                     -
content            -                                microk8s:microk8s     -
docker-support     microk8s:docker-privileged       :docker-support       -
docker-support     microk8s:docker-unprivileged     :docker-support       -
firewall-control   microk8s:firewall-control        :firewall-control     -
fuse-support       microk8s:fuse-support            -                     -
hardware-observe   microk8s:hardware-observe        :hardware-observe     -
home               microk8s:home                    :home                 -
home               microk8s:home-read-all           :home                 -
kernel-crypto-api  microk8s:kernel-crypto-api       -                     -
```

Figure 15.12 – MicroK8s snap interfaces

In the end, all snaps would need to aim for a strict confinement level, using only the APIs required for the application to run properly, and nothing else. Additionally, Ubuntu Core necessitates that snaps are on a strict confinement model.

Read more about interfaces and confinement in the Snapcraft documentation at https://docs.snapcraft.io.

Now that we have installed MicroK8s, let's verify whether the node status is Ready using the kubectl get nodes command as follows:

```
kubectl get nodes
```

If the installation is successful, then you should see the following output:

```
ubuntu@controlplane:~$ kubectl get nodes
NAME            STATUS    ROLES     AGE    VERSION
controlplane    Ready     <none>    12m    v1.23.3-2+0d2db09fa6fbbb
ubuntu@controlplane:~$ 
```

Figure 15.13 – Verifying whether the node is in a Ready state

Since MicroK8s is packaged as a snap, it will automatically upgrade to newer point releases. Also, the strictly confined MicroK8s version of the snap is currently on a dedicated snap channel that is synchronized with the latest version of upstream of Kubernetes, that is, an open source version of Kubernetes managed and maintained by the Cloud Native Computing Foundation.

Based on MicroK8s releases, channels are formed of a track (or series) and an anticipated level of stability (*stable*, *candidate*, *beta*, and *edge*). For more information about releases and channels, run the following command:

```
snap info microk8s
```

The following is the output after the execution of the preceding command, which shows the list of channels (*stable*, *candidate*, *beta*, and *edge*) and their release dates:

```
channels:
  1.24/stable:      v1.24.0  2022-05-13 (3272) 230MB classic
  1.24/candidate:   v1.24.0  2022-05-13 (3272) 230MB classic
  1.24/beta:        v1.24.0  2022-05-13 (3272) 230MB classic
  1.24/edge:        v1.24.2  2022-06-20 (3475) 230MB classic
  latest/stable:    v1.24.0  2022-05-13 (3272) 230MB classic
  latest/candidate: v1.24.0  2022-05-13 (3273) 230MB classic
  latest/beta:      v1.24.0  2022-05-13 (3273) 230MB classic
  latest/edge:      v1.24.2  2022-06-21 (3479) 230MB classic
  dqlite/stable:    -
  dqlite/candidate: -
```

Figure 15.14 – MicroK8s list of channels (stable, candidate, beta, and edge) and the release dates

Repeat the MicroK8s installation process on the other nodes as well.

The output of the command for the MicroK8s installation on the worker node is as follows:

```
ubuntu@worker1:~$ sudo snap install microk8s --channel=latest/edge/strict
microk8s (edge/strict) v1.24.1 from Canonical√ installed
ubuntu@worker1:~$
```

Figure 15.15 – Successful MicroK8s snap installation on the worker1 node

The following `microk8s status` command execution output confirms that MicroK8s is running successfully on the worker node as well:

```
ubuntu@worker1:~$ microk8s status
microk8s is running
high-availability: no
  datastore master nodes: 127.0.0.1:19001
  datastore standby nodes: none
addons:
  enabled:
    ha-cluster              # Configure high av
  disabled:
    dashboard               # The Kubernetes da
    dashboard-ingress       # Ingress definitio
```

Figure 15.16 – Verifying whether MicroK8s is running

Now that MicroK8s is running, the following step is to check whether the `kubectl get nodes` command displays the node in a `Ready` state as indicated in the command execution output as follows:

```
ubuntu@worker1:~$ kubectl get nodes
NAME       STATUS    ROLES     AGE    VERSION
worker1    Ready     <none>    13m    v1.23.3-2+0d2db09fa6fbbb
ubuntu@worker1:~$
```

Figure 15.17 – Verifying whether the node is in a Ready state

We have completed the installation of MicroK8s on all boards. The following step is to add the worker node to the control plane node. Open the PuTTY shell to the control plane node and run the following command to generate the connection string:

```
sudo microk8s.add-node
```

The following is the output after the execution of the preceding command. It validates that the command was successfully executed and provides instructions for generating the connection string:

```
ubuntu@controlplane:~$ sudo microk8s.add-node
From the node you wish to join to this cluster, run the following:
microk8s join 192.168.1.7:25000/fba12c2f1bce9fbe70208443565aaa04/3e2f115c73d6

Use the '--worker' flag to join a node as a worker not running the control plane,
microk8s join 192.168.1.7:25000/fba12c2f1bce9fbe70208443565aaa04/3e2f115c73d6 --wo

If the node you are adding is not reachable through the default interface you can
he following:
microk8s join 192.168.1.7:25000/fba12c2f1bce9fbe70208443565aaa04/3e2f115c73d6
ubuntu@controlplane:~$
```

Figure 15.18 – Generating the connection string for adding nodes

As indicated by the preceding command execution output, the connection string is generated in the form of `<control plane_ip>:<port>/<token>`.

Adding the worker node

We now have the connection string to join with the control plane node. Open the PuTTY shell to the worker node and run the `join` command to add it to the cluster, as can be seen in the following command:

```
microk8s join <control plane_ip>:<port>/<token>
```

The command was successfully executed, and the node has joined the cluster, as shown in the following output:

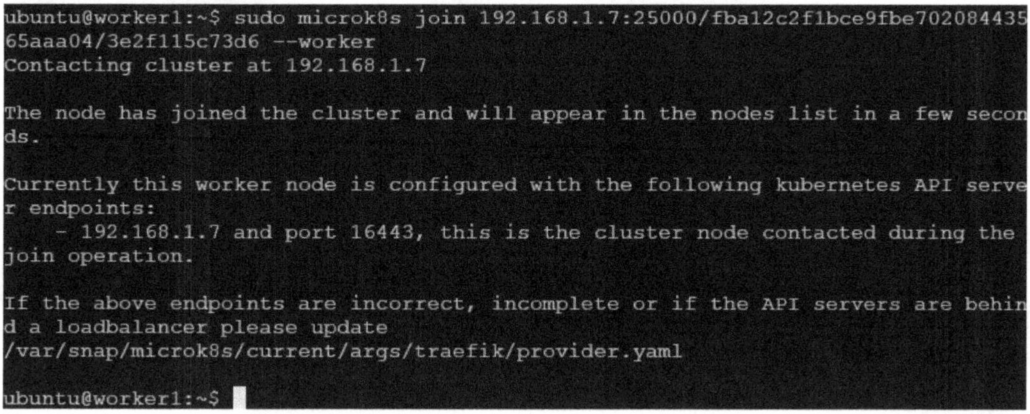

```
ubuntu@worker1:~$ sudo microk8s join 192.168.1.7:25000/fba12c2f1bce9fbe702084435
65aaa04/3e2f115c73d6 --worker
Contacting cluster at 192.168.1.7

The node has joined the cluster and will appear in the nodes list in a few secon
ds.

Currently this worker node is configured with the following kubernetes API serve
r endpoints:
    - 192.168.1.7 and port 16443, this is the cluster node contacted during the
join operation.

If the above endpoints are incorrect, incomplete or if the API servers are behin
d a loadbalancer please update
/var/snap/microk8s/current/args/traefik/provider.yaml

ubuntu@worker1:~$
```

Figure 15.19 – Add worker#1 node to the cluster

As indicated by the command execution output shown in *Figure 15.19*, you should be able to see the new node in a few seconds on the control plane node.

Use the following command to verify whether the new node is added to the cluster:

```
kubectl get nodes
```

The following command execution output shows that the control plane and `worker1` nodes are part of the cluster:

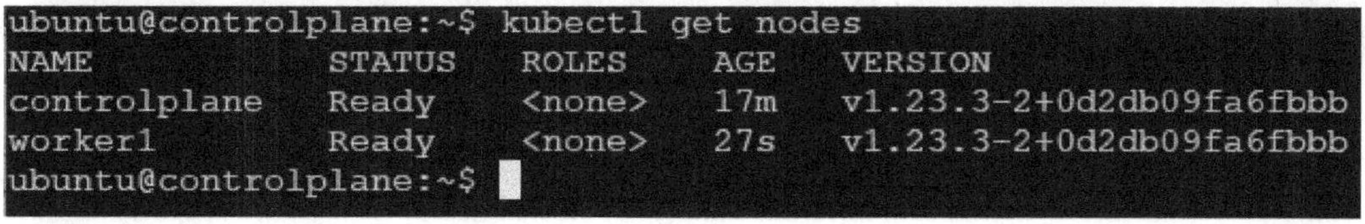

```
ubuntu@controlplane:~$ kubectl get nodes
NAME            STATUS    ROLES     AGE    VERSION
controlplane    Ready     <none>    17m    v1.23.3-2+0d2db09fa6fbbb
worker1         Ready     <none>    27s    v1.23.3-2+0d2db09fa6fbbb
ubuntu@controlplane:~$
```

Figure 15.20 – The cluster is ready and the control plane and worker1 nodes are part of the cluster

At this point, you have a fully functional multi-node Kubernetes cluster with strict confinement enabled. To summarize, we have installed the MicroK8s snap on the Raspberry Pi boards running Ubuntu Core and joined multiple deployments to form the cluster. We've seen how to add nodes to the cluster as well. In the following section, we are going to deploy a sample application on the MicroK8s cluster that we just created.

Deploying a sample containerized application

In this section, we will be deploying one of the nginx deployments from the Kubernetes examples repository on our MicroK8s cluster setup, as follows:

```
kubectl apply -f https://k8s.io/examples/controllers/nginx-
deployment.yaml
```

The following is the output after the execution of the preceding command. It indicates that there is no error in the deployment, and in the following steps, we can verify whether the deployment is successful using the get pods command as follows:

```
ubuntu@controlplane:~$ kubectl apply -f https://k8s.io/examples/controllers/nginx-deployment.yaml
deployment.apps/nginx-deployment created
ubuntu@controlplane:~$
```

Figure 15.21 – Sample application deployment

Check the status of the pods to verify whether the application has been deployed and is running as follows:

```
kubectl get pods -l app=nginx
```

The following is the output after the execution of the preceding command, which indicates that pods are created and that their status is Running:

```
ubuntu@controlplane:~$ kubectl get pods -l app=nginx
NAME                               READY   STATUS    RESTARTS   AGE
nginx-deployment-57d554699f-clxd5  1/1     Running   0          3m9s
nginx-deployment-57d554699f-8hjbv  1/1     Running   0          3m9s
ubuntu@controlplane:~$
```

Figure 15.22 – Checking whether pods have a Running status

Great! We have just deployed and examined our sample application deployment on the Raspberry multi-node cluster running Ubuntu Core.

To summarize, embedded Linux development utilizing snaps, Snapcraft, and Ubuntu Core is a lot quicker, safer, and more reliable than the current options available in the market. As it can package, distribute, and update any app through the global Snap Store, Snapcraft makes it simpler to find new software for your embedded devices. Additionally, the application updates either completely succeed or are not deployed at all. During both application and system updates, your embedded device running Ubuntu Core stays fully functional.

Summary

In this chapter, we learned how to install the MicroK8s snap with the strict confinement option, monitored the installation's progress, and managed the Kubernetes cluster running on Ubuntu Core. We also deployed a sample application and examined whether the application is able to run on a strict confinement-enabled Kubernetes cluster.

We also introduced a new embedded OS, Ubuntu Core, which complies with enterprise standards by enabling automated updates, app stores, and software management. We also learned that it is built from the ground up to be the most secure platform for connected devices. Furthermore, Ubuntu Core provides a modular design based on snaps, bullet-proof application updates, a seamless developer experience via Snapcraft, and built-in security to handle the challenges of embedded Linux development.

In this and earlier chapters, we have covered most of the implementation aspects that are required for your IoT/Edge computing applications using MicroK8s in detail; this includes running your applications on a multi-node Raspberry Pi cluster, configuring load balancing mechanisms, installing/configuring different CNI plugins for network connectivity, configuring logging, monitoring, and alerting options for your cluster, and building/deploying machine learning models and serverless applications.

Additionally, we have looked at setting up storage replication for your stateful applications, implementing a service mesh for your cross-cutting concerns, setting up a high-availability cluster to withstand component failure and continue to serve workloads without interruption, configuring containers with workload isolation, and running secured containers with isolation from a host system. In the following chapter, we'll look at how MicroK8s is uniquely positioned for accelerating IoT and Edge deployments and also key trends that are shaping up our new future.

16

Diving into the Future

According to a recent CNCF survey report (`https://www.cncf.io/reports/cncf-annual-survey-2021/`), 96% of enterprises use or are considering utilizing Kubernetes. Containers in general, and Kubernetes in particular, appear to be used less as the technology matures. Organizations appear to be employing serverless and managed services more intensely than in the past, and users no longer need to know about or understand the underlying container technology.

The industry has seen an exponential increase in the use of cloud-native technology over recent years. Modernizing applications with Kubernetes and containers has been a common theme for many businesses. The de facto DevOps standard for established businesses and start-ups is **continuous integration/ continuous deployment (CI/CD)** based on containers. The ideal platform for executing workloads at the edge is Kubernetes. Additionally, it has evolved into a hybrid computing platform that enables public cloud providers to operate their managed services in clusters set up in on-premises settings.

Edge-based infrastructure presents a myriad of challenges in terms of managing resources and workloads. In a shorter period of time, thousands of edge nodes and remote edge nodes would need to be controlled. The edge architecture of organizations is made to offer more centralized independence from the cloud, high-security requirements, and minimal latency.

Throughout this book, we have covered the following implementation aspects that address IoT/Edge computing scenarios using MicroK8s:

- Getting your Kubernetes cluster up and running

- Enabling core Kubernetes add-ons such as DNS and dashboards

- Creating, scaling, and performing rolling updates on multi-node Kubernetes clusters

- Working with various container networking options for networking – Calico/Flannel/Cilium

- Setting up MetalLB, and Ingress options for load balancing

- Using OpenEBS storage replication for stateful applications

- Configuring Kubeflow and running AI/ML use cases

- Configuring service mesh integration with Istio/Linkerd

- Running serverless applications using Knative and OpenFaaS

- Configuring logging/monitoring options (Prometheus, Grafana, Elastic, Fluentd, and Kibana)

- Configuring multi-node highly available Kubernetes clusters

- Configuring Kata Containers for secured containers

- Configuring strict confinement for running in isolation

Furthermore, we discussed the guidelines and best practices for designing and effectively implementing Kubernetes for your edge workloads in each chapter.

The importance of Kubernetes, the edge, and the cloud collaborating to drive sensible business decisions is becoming more and more evident as firms embrace digital transformation, Industry 4.0, industrial automation, smart manufacturing, and other advanced use cases.

Businesses that are transitioning to become digital-first enterprises increasingly rely on Kubernetes. Kubernetes is clearly the preferred platform for Edge computing, at least for those edges that require dynamic orchestration for apps and centralized administration of workloads. By enabling flexible and automated administration of applications over a disaggregated cloud environment, Kubernetes extends the advantages of cloud-native computing software development to the edge. In this final chapter, we're going to cover the following main topics:

- How MicroK8s is uniquely positioned for accelerating IoT and Edge deployments

- Looking forward – Kubernetes trends and industry outlook

How MicroK8s is uniquely positioned for accelerating IoT and Edge deployments

Edge gateways must efficiently utilize computational resources while dealing with a variety of protocols, including Bluetooth, Wi-Fi, 3G, 4G, and 5G. It is challenging to operate Kubernetes directly on edge servers because edge gateways have constrained computational capabilities. Some of the problems include the following:

- For better monitoring and management, separating the control plane and worker nodes from the edge and transferring the control plane to the cloud, where the control plane and worker nodes take the workload.

- Separating the cluster data store to handle heavy loads.

- Making worker nodes specifically for incoming and outgoing traffic will improve traffic management.

These problems will result in the development of several clusters, making the management of the entire infrastructure more challenging.

MicroK8s comes to the rescue, as it serves as a bridge between edge clusters and mainstream Kubernetes. Running with limited resources necessitates a small footprint, and full-fledged cloud resource pools can be orchestrated. We have seen in the earlier chapters that MicroK8s leverages immutable containers in Kubernetes for improved security and simpler operations. It aids in the creation of self-healing, high-availability clusters that select the best nodes for the Kubernetes data store automatically. When one of the cluster database nodes fails, another node gets promoted without the requirement for an administrator. MicroK8s is easy to install and upgrade, and it has robust security, making it ideal for micro clouds and Edge computing.

Some of the notable challenges in operating IoT edge

In this section, we will look at some of the significant problems associated with IoT edge operations:

- **Computation and resource constraints**: IoT edge devices' CPU and memory resources are usually constrained, therefore, they must be utilized wisely and maintained for the solution's mission-critical functionality.

- **Remote and resource management**: A manual method for deploying, administering, and maintaining devices will be difficult and time-consuming when the cluster or edge network expands quickly. Some of the prominent issues are as follows:

  - Using device resources efficiently, including CPU, memory, networking, and edge-device I/O ports, as well as their remote monitoring and management

  - Controlling CPU cores and co-processing (for example, GPU) to specific workloads, as well as hosting and scaling any mix of apps

  - Updates that are automated, remote, and have the ability to roll back in order to avoid bricking of the devices

  - Easy migration to different backends and automated connection to one or more of the backends (such as the cloud or on-premises infrastructure)

  - A distributed, secure firewall that securely routes data over networks in accordance with the policies defined

- **Security and trust**: The IoT edge devices must be protected from unauthorized access. High-scale environments pose serious challenges for device anonymity and traceability, as well as discovery, authentication, and trust building at the IoT edge. To guarantee that several IoT apps run in isolation from one another in the device, an extra security layer is a critical mandate.

- **Reliability and fault tolerance**: Self-managing and self-configuring solutions are needed on the edge network due to the volume of IoT devices in the system. IoT apps need to have the ability to fix any problems that develop throughout the course of their existence. Some of the frequent requirements in the IoT edge include resilience to failures and mitigating denial of service attacks.

- **Scalability**: In the IoT ecosystem, sensors or actuators are increasingly in charge of everything. Both the volume and the number of data collection points are growing quickly. It is normal for hundreds of new sensors or actuators to be added in a short amount of time while the IoT environment is still operating in many applications (such as smart city and smart traffic systems). As a result, the requirement to scale the IoT ecosystem and data management is critical. Additionally, edge-based services are challenged by costs as well as other factors, including workload monitoring, storage capacity, dynamic resource allocation, and data transfer rate.

- **Scheduling and load balancing**: To sustain massive systems where data is shared over several services, edge computing is totally dependent on load balancing and scheduling methods. It is necessary to make data, software, and infrastructure available at a lower cost in a safe, dependable, and adaptable way in order to assure optimal usage of computational resources. Additionally, a reliable system for scheduling and load balancing is required.

Now that we've seen the major difficulties in managing IoT edge infrastructure, we'll examine how MicroK8s Kubernetes is effective in resolving those challenges.

How MicroK8s Kubernetes is benefiting edge devices

MicroK8s is well positioned for expediting IoT and edge deployments due to its ability to improve Kubernetes' productivity and reduce complexity. In this section, we will look at how MicroK8s Kubernetes is benefiting edge devices:

- **Scalability**: For many IoT solutions, scalability is the main concern. An infrastructure that can independently scale horizontally or vertically is necessary to support additional devices and process terabytes of data in real time. Compared to conventional virtual machines, containers can be generated faster since they are lightweight. One of MicroK8s Kubernetes' primary advantages is its simplicity in scaling across network clusters, independence in scaling containers, and ability to restart automatically without affecting those services.

- **High availability**: For IoT solutions to conduct crucial business functions, edge devices must be readily available and trustworthy. Due to the fact that each container has its own IP address, it is simple to distribute loads among them and restart applications when a container stops functioning. We have seen various examples of how to use the load balancing functionality and run multiple replicas for high availability. Also, we have looked at steps to set up an HA cluster to withstand component failures.

- **Efficient use of resources**: Due to its effective resource management, Kubernetes reduces the cost of hosting IoT applications. MicroK8s is the compact, optimized version of Kubernetes, which offers a layer of abstraction on top of hosted virtual machines, bare metal instances, or on the cloud. Administrators can focus on spreading out application service deployment across the most infrastructure possible, which lowers the overall cost of running infrastructure for an IoT application.

- **Deployment to the IoT Edge**: Deploying software updates to edge devices without disrupting services is a significant IoT challenge. Microservices that gradually roll out updates to services can be run via Kubernetes. A rolling update strategy is typically used in Kubernetes installations to roll out updates to pod versions. By leaving certain instances operating (such as Pod Disruption Budgets) at any given time while the updates are being made, it is possible to achieve zero service downtime. Old pods are only evicted after the new deployment version's traffic-ready pods are enabled and ready to replace them. As a result, applications can be scaled horizontally or upward with a single command.

- **Enabling DevOps for IoT**: To meet consumer needs, IoT solutions must be updated smoothly with no user downtime. Development teams can efficiently verify, roll out, and deploy changes to IoT services with the aid of CI/CD tools that are available for Kubernetes. Additionally, Kubernetes is supported by a number of cloud service providers, including Azure, Google Cloud, and AWS. As a result, switching to any cloud service will be simple in the future.

IoT-dependent industries are concentrating on implementing mission-critical services in edge devices to increase the responsiveness of solutions and lower costs. Solutions that are getting built on the Kubernetes platforms offer a standard framework for implementing IoT services at the edge. Continuous advancements from the Kubernetes community make it possible to build IoT solutions that are scalable, reliable, and deployable in a distributed environment.

In the next section, we will look at some of the trends that are driving Kubernetes and its adoption.

Looking forward – Kubernetes trends and industry outlook

According to a Gartner report titled *Emerging Technologies: Kubernetes and the Battle for Cloud-Native Infrastructure, October 2021*, "*By 2025, 85% of organizations will run containers in production, up from less than 30% in 2020.*" In this section, we will look at some of the key trends that are going to drive Kubernetes adoption and use in enterprises.

Trend 1 – security is still everyone's concern

Significant security concerns are posed by containers and Kubernetes that are already well known. In the last 12 months, 93% of Kubernetes environments suffered at least one security incident. This is likely due to a number of problems, such as a lack of security expertise about containers and Kubernetes, insufficient or unsuitable security tooling, and central security teams that are unable to keep up with rapidly developing application development teams who consider security to be an afterthought.

An application's security posture can be affected by several configuration options in Kubernetes. There could be exposures due to misconfigurations in the container and Kubernetes environments. Businesses now know that they cannot adequately secure containerized environments if security is not incorporated into every phase of their development life cycle. DevSecOps methodology is now becoming an integral component of managing containerized environments.

I have highlighted the need for DevSecOps in the recent *Kubernetes and cloud-native operations report, 2022*. Read more on the analysis and takeaways here: `https://juju.is/cloud-native-kubernetes-usage-report-2022#key-takeaways`.

Another aspect that could be affected is the software industry's supply chain. The process of creating modern software involves combining and merging multiple parts that are freely accessible as open source projects. A vulnerable software component could seriously harm other parts of the application and entire deployments as well as the intricate software supply chain. In the following days, new initiatives, projects, and so on may be introduced to safeguard the software supply chain.

The next breakthrough is the **extended Berkeley Packet Filter** (**eBPF**), which gives cloud-native developers the flexibility to create components for secure networking, service mesh, and observability. We have seen an example of using eBPF with Cilium in *Chapter 6, Configuring Connectivity for Containers*. In the coming days, eBPF could become prevalent in the security and networking space.

Trend 2 – GitOps for continuous deployment

GitOps provides well-known Git-based processes and is a crucial tool since it enables quick rollbacks and can be used as a single source of truth for state reconciliation.

Natively, there are many ways to integrate GitOps, including Flux CD, Argo CD, Google Anthos Config Management, Codefresh, and Weaveworks.

Tens of thousands of Kubernetes clusters running at the edge or in hybrid settings may now be easily managed using GitOps' support for multitenant and multicluster deployments.

GitOps is thus rising to the top as the preferred method for continuous deployment. In the upcoming days, GitOps is going to become the gold standard for running and deploying Kubernetes apps and clusters.

Trend 3 – App store for operators

Without requiring any additional technical expertise, Kubernetes can scale and manage stateless applications, including web apps, mobile backends, and API services. Kubernetes' built-in capabilities handle these tasks simply.

However, stateful applications, such as databases and monitoring systems, necessitate extra domain expertise that Kubernetes lacks. To scale, update, and reconfigure these applications requires an extra level of understanding of the applications that are deployed. To manage and automate the life cycle of an application, Kubernetes operators include this unique domain knowledge in their extensions.

Kubernetes operators make these procedures scalable, repeatable, and standardized by eliminating laborious manual application administration duties.

Operators make it simpler for application developers to deploy and maintain the supporting services needed by their apps. Additionally, they offer a standardized method for distributing applications on Kubernetes clusters and lessen the requirements for support by spotting and fixing application issues for infrastructure engineers and vendors. We have seen an example of an operator pattern in *Chapter 8, Monitoring the Health of Infrastructure and Application*, where we deployed the Prometheus Operator for Kubernetes, which handles simplified monitoring definitions for Kubernetes services as well as Prometheus instance deployment and management.

However, there is concern surrounding the "*true*" origin and accessibility of operators in order to alleviate the fundamental worries of organizations adopting new technologies, particularly open source solutions.

Like Charmhub.io (`https://charmhub.io/`), there should be a central place such as an app store where people can publish and consume operators. There will be specific ownership of the artifacts, validation, and different flavors of them. And the "store" will have enough information for people to choose the right flavor, based on documentation, ratings, the different publishers, and so on.

I have outlined the *app store for operators* idea in the recent *Kubernetes and cloud-native operations report, 2022*. Read more about the analysis and takeaways here: `https://juju.is/cloud-native-kubernetes-usage-report-2022#key-takeaways`.

Trend 4 – Serverless computing and containers

Analysts at Gartner anticipated the growth of serverless computing, or **function-as-a-service (FaaS)**, much earlier (`https://blogs.gartner.com/tony-iams/containers-serverless-computing-pave-way-cloud-native-infrastructure/`).

Imagine that you have a complicated containerized system that is executing shared services (such as integration, database operations, and authentication) that are triggered by events. To offload complexity from your containerized setup, you can separate such duties into a serverless function rather than running them in a container.

Additionally, a serverless application can be readily expanded using containers. In most scenarios, serverless functions save data; you may integrate and communicate stateful data between serverless and container architectures by mounting these services as Kubernetes Persistent Volumes.

Kubernetes-based Event Driven Autoscaler (KEDA) (`https://keda.sh/`) comes to the rescue for running event-driven Kubernetes workloads, such as containerized functions, as it provides fine-grained autoscaling. Functions' runtimes receive event-driven scaling functionality from KEDA. Based on the load, KEDA can scale from *zero* instances (*when no events are happening*) out to *n* instances. By making custom metrics available to the Kubernetes autoscaler (*Horizontal Pod Autoscaler*), it enables autoscaling. Any Kubernetes cluster can replicate serverless function capabilities by utilizing functions, containers, and KEDA.

Knative (`https://knative.dev/docs/`) is another framework that integrates scaling, the Kubernetes Deployment model, and event and network routing. Through a Knative-service resource, the Knative platform, which is built on top of Kubernetes, adopts an opinionated stance on workload management. CloudEvents is the foundation of Knative, and Knative services are essentially functions that are triggered and scaled by events, whether they are CloudEvents events or straightforward HTTP requests. Knative scales quickly in response to changes in event rates because it uses a pod sidecar to monitor event rates. Additionally, Knative offers scaling to zero, enabling more precise workload scaling, ideal for microservices and functions.

Traditional Kubernetes Deployments/Services are used to implement Knative services, and changes to Knative services (such as adding a new container image) generate simultaneous Kubernetes Deployment/Service resources. With the routing of HTTP traffic being a part of the Knative service resource description, this is used by Knative to implement the blue/green and canary deployment patterns.

As a result, while designing the deployment of an application on Kubernetes, developers should use the Knative service resource and its associated resources for specifying event routing. Using Knative means developers will primarily be concerned with the Knative service, and Deployments are handled by the Knative platform, similar to how we frequently deal with Kubernetes today through deployment resources and let Kubernetes manage Pods.

OpenFaaS is a framework for creating serverless functions using the Docker and Kubernetes container technologies. Any process can be packaged as a function, allowing it to consume a variety of web events without having to write boilerplate code over and over. It is an open source initiative that is gaining a lot of traction in the community.

I have covered the OpenFaaS framework in my blog: `https://www.upnxtblog.com/index.php/2018/10/19/openfaas-tutorial-build-and-deploy-serverless-java-functions/`.

In *Chapter 10, Going Serverless with Knative and OpenFaaS Frameworks*, we have looked at how to deploy samples on Knative and OpenFaaS platforms and used their endpoints to invoke them via the CLI. We also looked at how serverless frameworks scale down pods to zero when there are no requests and spin up new pods when there are more requests. We've also discussed some guiding principles to keep in mind when developing and deploying serverless applications.

In the following days, there could be new initiatives and open source projects launched that could foster innovation on running serverless and containers.

Trend 5 – AI/ML and data platforms

For workloads including **machine learning** and **artificial intelligence** (**ML** and **AI**), Kubernetes has been widely used. Organizations have experimented with a variety of techniques to deliver these capabilities, including manual scaling on bare metal, VM scaling on public cloud infrastructure, and **high-performance computing** (**HPC**) systems. However, AI algorithms frequently demand significant computational power.

Kubernetes may be the most effective and straightforward choice. The ability to package AI/ML workloads as containers and run them as clusters on Kubernetes gives AI projects flexibility, maximizes resource usage, and gives data scientists a self-service environment.

Without having to adjust GPU support for each workload, containers let data science teams build and reliably reproduce validated setups. NVIDIA and AMD have added experimental GPU support to the most recent version of Kubernetes. Additionally, NVIDIA offers a library of preloaded containers and GPU-optimized containerized ML applications (`https://developer.nvidia.com/ai-hpc-containers`).

In *Chapter 9, Using Kubeflow to Run AI/MLOps Workloads*, we went over how to set up an ML pipeline that will develop and deploy an example model using the Kubeflow ML platform. We also noticed that Kubeflow on MicroK8s is easy to set up and configure, as well as lightweight and capable of simulating real-world conditions while constructing, migrating, and deploying pipelines.

We can expect to see more and more AI/ML and data platforms moving toward Kubernetes.

Trend 6 – Stateful applications

Today, stateful applications are the norm. While technology innovations such as containers and microservices have simplified the development of cloud-based systems, their agility has made managing stateful processes more difficult.

Stateful apps must be executed in containers more frequently. In complicated contexts such as the edge, public cloud, and hybrid cloud, containerized apps can streamline deployment and operations. For **CI/CD** to provide a seamless transition from development to production, maintaining the state is equally crucial.

Kubernetes has made significant improvements to facilitate running stateful workloads by giving platform administrators and application developers the necessary abstractions. The abstractions ensure that different types of file and block storage are available wherever a container is scheduled.

In *Chapter 11, Managing Storage Replication with OpenEBS*, we looked at how to configure and implement a PostgreSQL stateful workload utilizing the OpenEBS storage engine. We also looked at some Kubernetes storage best practices, as well as guidelines for choosing data engines.

To summarize, we can expect to see a strong trend toward automated security and continuous compliance for container and Kubernetes infrastructure in 2022, as well as the development of best practices. This will be especially important for businesses that must adhere to strict compliance standards.

Summary

To summarize, the importance of Kubernetes, the edge, and the cloud collaborating to drive sensible business decisions is becoming more and more evident as firms embrace digital transformation, Industry 4.0, industrial automation, smart manufacturing, and all the advanced use cases. We've also explored different deployment approaches that demonstrate how Kubernetes may be utilized to run edge workloads. Throughout this book, we have covered the majority of the implementation aspects of IoT/Edge computing applications using MicroK8s. Kubernetes is clearly the preferred platform for Edge computing.

We have also seen how MicroK8s is uniquely positioned for accelerating IoT and Edge deployments. Furthermore, we have looked at some of the key trends that are going to shape the future.

Congrats! You have successfully completed this book. As you continue on your Kubernetes journey, I'm confident that you would have benefited from the examples, scenarios, use cases, best practices, and recommendations that we discussed throughout this book.

In conclusion, Kubernetes, with its rapidly expanding ecosystem and variety of tools, support, and services, is quickly becoming a helpful tool, particularly as more organizations shift to the cloud.

According to Canonical's 2022 *Kubernetes and Cloud Native Operations Survey* (`https://juju.is/cloud-native-kubernetes-usage-report-2022`), 48% of respondents indicate the biggest barriers to migrating to or using Kubernetes and containers are a lack of in-house capabilities and limited staff.

As indicated in the report, there is a skill deficit as well as a knowledge gap, which I believe this book can solve by covering crucial areas that are required to bring you up to speed in no time.

To keep up with updates, you may subscribe to my blog at `https://www.upnxtblog.com`.

I also look forward to hearing about your experiences, opinions, and suggestions on `https://twitter.com/karthi4india`.

The following are some excellent MicroK8s resources to support you on your journey.

Further reading

- Official MicroK8s documentation: `https://microk8s.io/docs`.

- MicroK8s tutorials: `https://microk8s.io/docs/tutorials`.

- MicroK8s command reference: `https://microk8s.io/docs/command-reference`.

- Services and ports used: `https://microk8s.io/docs/services-and-ports`.

- If you are unable to resolve your problem and feel it is due to a bug in MicroK8s, please submit an issue to the project repository: `https://github.com/ubuntu/microk8s/issues/`.

- Contributing to MicroK8s documentation: `https://microk8s.io/docs/docs`.

- Contributing to MicroK8s: `https://github.com/canonical/microk8s/blob/master/docs/build.md`.

Frequently Asked Questions About MicroK8s

The following FAQs are not exhaustive, but they are important for running your Kubernetes cluster:

1. How do I find out what the status of a deployment is?

    Use the `kubectl get deployment <deployment>` command. If `DESIRED`, `CURRENT`, and `UP-TO-DATE` are all equal, then the deployment has succeeded.

2. How do I troubleshoot a pod with a `Pending` status?

    A pod with the Pending status cannot be scheduled onto a node. Inspecting the pod using `kubectl describe pod <pod>` will give you details on why the pod is stuck. Additionally, you can use the `kubectl logs <pod>` command to understand if there is contention.

    The most common reason for this issue is some pod requesting more resources.

3. How do I troubleshoot a `ContainerCreating` pod?

    Unlike a `Pending` pod, a ContainerCreating pod is scheduled onto the node but due to some other reason, it cannot start up properly. Using `kubectl describe pod <pod>` will give you details on why the pod is stuck on the ContainerCreating status.

    The most common reasons for the above issue include CNI errors from being started up properly. There could also be errors due to volume mount failures.

4. How do I troubleshoot a pod with a `CrashLoopBackoff` status?

    When a pod fails due to an error, this is the standard error message. The `kubectl logs <pod>` command would usually show the error messages from the recent execution. From those messages, you can find out what caused the issue and resolve it.

    If the container is still running, you can use the `kubectl exec -it <pod> -- bash` command to enter the container shell and then debug it.

5. How do I roll back a particular deployment?

    If you use the `-record` parameter along with the `kubectl apply` command, Kubernetes stores the previous deployment in its history. You can then use `kubectl rollout history deployment <deployment>` to show prior deployments.

    The last deployment can be restored using the `kubectl rollout undo deployment <deployment>` command.

6. How do I force a pod to run on specific nodes?

    Some of the common methods that are used are the `nodeSelector` field and affinity and anti-affinity.

    The simplest recommendation is to use a node selection constraint with the `nodeSelector` field in your pod definition to define which node labels the target node should have. Kubernetes uses that information to schedule only the nodes with the labels you specify.

7. How do I force replicas to distribute on different nodes?

    Kubernetes attempts node anti-affinity by default, but this is not a hard requirement; it is by best effort, but it will schedule many pods on the same node if that is the only option available. You can learn more about node selection here: `http://kubernetes.io/docs/user-guide/node-selection/`.

8. How do I list all the pods on a node?

    Use the following command:

    ```
    kubectl get pods -A  --field-selector spec.nodeName=<node
    name> | awk '{print $2"  "$4}'
    ```

    A more detailed kubectl cheat sheet can be found at `https://kubernetes.io/docs/reference/kubectl/cheatsheet/`.

9. How do I monitor a pod that is always running?

    To do this, you can make use of the liveness probe feature.

    A liveness probe always checks whether an application in a pod is running, and if this check fails, the container gets restarted. This is useful in many situations where the container is running but the application inside it crashes.

    The following code snippet demonstrates the liveness probe feature:

    ```
    spec:
    containers:
    - name: liveness
    image: k8s.gcr.io/liveness
    args:
    - /server
    livenessProbe:
          httpGet:
              path: /healthcheck
    ```

10. What is the difference between replication controllers and replica sets?

The selectors are the sole distinction between replication controllers and replica sets. Replication controllers are no longer supported in the most recent version of Kubernetes, and their specifications don't mention selectors either.

More details can be found at `https://Kubernetes.io/docs/concepts/workloads/controllers/replicaset/`.

11. What is the role of `kube-proxy`?

The following are the roles and responsibilities of `kube-proxy`:

- For every service, it assigns a random port to the node it's running on and assigns a proxy to the service.

- Installs and maintains iptable rules that intercept incoming connections to a virtual IP and port and also routes them to the port.

The kube-proxy component oversees host subnetting and makes services available to other components. Since kube-proxy manages network communication, shutting down the control plane does not prevent a node from handling traffic. It operates similarly to a service. The connection will be forwarded by iptables to kube-proxy, which will then use a proxy to connect to one of the service's pods. Whatever is in the endpoints is routed through kube-proxy using the target address.

12. How do I test the deployment manifest without executing it?

To test the manifest, use the `--dry-run` flag. This is extremely useful for determining whether the YAML syntax is appropriate for a specific Kubernetes object and also ensures that a spec contains the required key-value pairs:

```
kubectl create -f <test manifest.yaml> --dry-run
```

13. How do I package a Kubernetes application?

Helm is a package manager that allows users to package, configure, and deploy Kubernetes applications and services. You can learn more about Helm here: `https://helm.sh/`.

For a quick-start guide, please refer to `https://www.upnxtblog.com/index.php/2019/12/02/helm-3-0-0-is-outhere-is-what-has-changed/`.

14. What are `init` containers?

In Kubernetes, a pod can have several containers. The `init` container runs before any other containers in the pod.

The following is an example that defines a simple pod with two `init` containers. The first is waiting for `myservice`, while the second is waiting for `mydb`. When both `init` containers are finished, the pod executes the app container from its `spec:` section.

More details can be found here: `https://kubernetes.io/docs/concepts/workloads/pods/init-containers/`.

The following code snippet demonstrates how `initContainers` works:

```
apiVersion: v1
kind: Pod
metadata:
  name: sample-app-pod
  labels:
    app: sample-app
spec:
  containers:
  - name: sample-app-container
    image: busybox:1.28
    command: ['sh', '-c', 'echo The app is running! &&
sleep 3600']
  initContainers:
  - name: init-myservice
    image: busybox:1.28
    command: ['sh', '-c', "until nslookup myservice.$(cat
/var/run/secrets/kubernetes.io/serviceaccount/namespace).
svc.cluster.local; do echo waiting for myservice; sleep
2; done"]
  - name: init-mydb
    image: busybox:1.28
    command: ['sh', '-c', "until nslookup mydb.$(cat /
var/run/secrets/kubernetes.io/serviceaccount/namespace).
svc.cluster.local; do echo waiting for mydb; sleep 2;
done"]
```

15. How can I drain the pods from nodes for maintenance?

    Use the `drain` command, as follows:

    ```
    kubectl drain <node>
    ```

    When you execute the preceding command, it designates the node as unscheduled for newer pods and then evicts or deletes the existing pods.

    Once you have finished maintaining the node and you want to join the cluster, issue the `uncordon` command, as follows:

    ```
    kubectl uncordon <node>
    ```

More details can be found at `https://kubernetes.io/docs/tasks/administer-cluster/safely-drain-node/`.

16. What is a pod security policy?

Pod security policies in Kubernetes are configurations that govern which security features a pod has access to. They are a form of cluster-level resource that helps you control a pod's security.

More details can be found at `https://kubernetes.io/docs/concepts/security/pod-security-policy/`.

17. What is `ResourceQuota` and why do we need it?

The `ResourceQuota` object limits aggregate resource consumption per namespace. It can limit the number of objects that can be generated in a namespace by type, as well as the total amount of compute resources that resources in that project can consume.

More details can be found at `https://kubernetes.io/docs/concepts/policy/resource-quotas/`.

Index

Symbols

N

O

`Packt.com`

Subscribe to our online digital library for full access to over 7,000 books and videos, as well as industry leading tools to help you plan your personal development and advance your career. For more information, please visit our website.

Why subscribe?

- Spend less time learning and more time coding with practical eBooks and Videos from over 4,000 industry professionals

- Improve your learning with Skill Plans built especially for you

- Get a free eBook or video every month

- Fully searchable for easy access to vital information

- Copy and paste, print, and bookmark content

Did you know that Packt offers eBook versions of every book published, with PDF and ePub files available? You can upgrade to the eBook version at `packt.com` and as a print book customer, you are entitled to a discount on the eBook copy. Get in touch with us at `customercare@packtpub.com` for more details.

At `www.packt.com`, you can also read a collection of free technical articles, sign up for a range of free newsletters, and receive exclusive discounts and offers on Packt books and eBooks.

Other Books You May Enjoy

If you enjoyed this book, you may be interested in these other books by Packt:

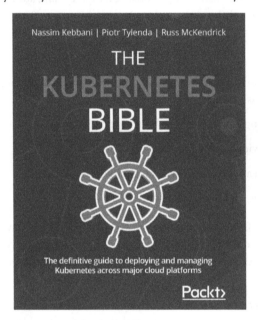

The Kubernetes Bible

Nassim Kebbani, Piotr Tylenda, Russ McKendrick

ISBN: 9781838827694

- Manage containerized applications with Kubernetes

- Understand Kubernetes architecture and the responsibilities of each component

- Set up Kubernetes on Amazon Elastic Kubernetes Service, Google Kubernetes Engine, and Microsoft Azure Kubernetes Service

- Deploy cloud applications such as Prometheus and Elasticsearch using Helm charts

- Discover advanced techniques for Pod scheduling and auto-scaling the cluster

- Understand possible approaches to traffic routing in Kubernetes

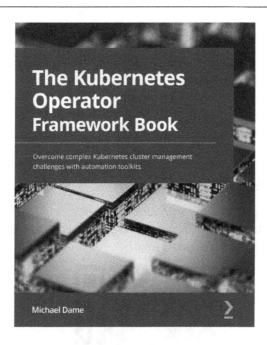

The Kubernetes Operator Framework Book

Michael Dame

ISBN: 9781803232850

- Gain insight into the Operator Framework and the benefits of operators
- Implement standard approaches for designing an operator
- Develop an operator in a stepwise manner using the Operator SDK
- Publish operators using distribution options such as OperatorHub.io
- Deploy operators using different Operator Lifecycle Manager options
- Discover how Kubernetes development standards relate to operators
- Apply knowledge learned from the case studies of real-world operators

Packt is searching for authors like you

If you're interested in becoming an author for Packt, please visit `authors.packtpub.com` and apply today. We have worked with thousands of developers and tech professionals, just like you, to help them share their insight with the global tech community. You can make a general application, apply for a specific hot topic that we are recruiting an author for, or submit your own idea.

Share Your Thoughts

Now you've finished *IoT Edge Computing with MicroK8s*, we'd love to hear your thoughts! Scan the QR code below to go straight to the Amazon review page for this book and share your feedback or leave a review on the site that you purchased it from.

https://packt.link/r/1803230630

Your review is important to us and the tech community and will help us make sure we're delivering excellent quality content.

www.ingramcontent.com/pod-product-compliance
Lightning Source LLC
Chambersburg PA
CBHW081504050326
40690CB00015B/2913